THE AGRICULTURAL REVOLUTION OF THE 20TH CENTURY

THE AGRICULTURAL REVOLUTION OF THE 20TH CENTURY

DON PAARLBERG AND PHILIP PAARLBERG

IOWA STATE UNIVERSITY PRESS / AMES

© 2000 Iowa State University Press
All rights reserved

Iowa State University Press
2121 South State Avenue, Ames, IA 50014

Orders: 1-800-862-6657
Office: 1-515-292-0140
Fax: 1-515-292-3348
Web site: www.isupress.com

Cover photo by Mark Fitch: Restored Case steam engine used to power old farm machinery such as a thresher.

Authorization to photocopy items for internal or personal use, or the internal or personal use of specific clients, is granted by Iowa State University Press, provided that the base fee of $.10 per copy is paid directly to the Copyright Clearance Center, 222 Rosewood Drive, Danvers, MA 01923. For those organizations that have been granted a photocopy license by CCC, a separate system of payments has been arranged. The fee code for users of the Transactional Reporting Service is 0-8138-2198-3/2000 (hardcover); 0-8138-0409-4/2001 (paperback) $.10.

♾ Printed on acid-free paper in the United States of America

First edition, 2000
First paperback edition, 2001

Library of Congress Cataloging-in-Publication Data

Paarlberg, Don
    The agricultural revolution of the 20th century / Don Paarlberg and Philip Paarlberg.
      p. cm.
    Includes bibliographical references (p.   ) and index.
    ISBN 0-8138-2198-3 (hardcover); 0-8138-0409-4 (paperback) (alk. paper)
     1. Agricultural innovations. 2. Agriculture—Technology transfer. 3. Agriculture and state. I. Paarlberg, Don. II. Paarlberg Philip L. III. Title.

S494.5.I5 P22 2000
338.1'6—dc21   00-057230

The last digit is the print number: 9 8 7 6 5 4 3 2 1

Don Paarlberg is professor emeritus and former Hillenbrand Distinguished Professor in the Department of Agricultural Economics at Purdue University, West Lafayette, Indiana. Dr. Paarlberg's professional life has been devoted to teaching and research in agricultural economics and public service in the federal government. He has held appointments as assistant to the president, coordinator of the Food for Peace Program, assistant secretary of agriculture, director of agricultural economics in the U.S. Department of Agriculture, board member of the Commodity Credit Corporation, chairman of the U.S. Department of Agriculture Graduate School, and economic advisor to four secretaries of agriculture. Dr. Paarlberg is also the author of many articles and seven books.

Philip Paarlberg received his Ph.D. degree in agricultural economics from Purdue University, West Lafayette, Indiana. His thesis on the international wheat market received the outstanding Ph.D. thesis award from the American Agricultural Economics Association. While serving with the U.S. Department of Agriculture, he received a USDA Certificate of Merit, the Economic Research Service Administrator's Special Merit Award, and a USDA Superior Service Award for his contributions to policy analysis of world grain and oilseed markets and agricultural trade. Dr. Paarlberg teaches international trade theory and policy in the Department of Agricultural Economics at Purdue and conducts research on agricultural trade policies with a focus on export subsidies, sanitary and phytosanitary trade barriers, and the economic transformation of formerly centrally planned economies.

Dedication

This book is dedicated to the West Lafayette Kaffee Klatsch, a group of about 20 men, all retired, most of them veterans of the Agricultural Revolution, who meet daily at 8 a.m. at a local cafeteria to drink coffee and tell stories, some of which are true.

Contents

Preface, ix
Acknowledgments, xi
Introduction, xiii

Section I The Technological Revolution

Chapter 1   19th Century Agriculture, 3
Chapter 2   Mechanized Agriculture, 21
Chapter 3   Chemical Advances, 31
Chapter 4   Biological Changes, 43
Chapter 5   Information Explosion, 51
Chapter 6   Management: Integration of the Disciplines, 59
Chapter 7   The Agricultural Revolution in a Global Context, 69
Chapter 8   Exporting the Revolution, 99

Section II The Deeper Dimension

Chapter 9   Birth Pangs—And a Live Birth, 111
Chapter 10 Consumer Benefits, 117
Chapter 11 Political Issues Arising From the Agricultural Revolution, 127

Section III Looking Ahead

Chapter 12 Into the 21st Century, 137

References, 145
Other Readings, 147
Index, 149

Preface

The great changes in American agriculture that occurred during the 20th century have been well described by a large number of competent authors. The present writing is an attempt to bring together and interpret, in narrative and visual form, the meaning of these profound changes.

Some of the agricultural changes that occurred during the 20th century include the following:

From horse to tractor and from buggy to automobile

From the 10-horsepower to the more than 400-horsepower tractor

From hoe to herbicide

From genetic mythology to bioengineering

From lead arsenate to biodegradable chemicals

From a national average of 21 to 121 bushels of corn per acre

From husking hook to corn combine

From Sam and Henry, black-face comedians, on radio to Larry King live on television

From an old-time religion to a more open form of faith

From carrying water to turning the kitchen faucet

From backhouse to indoor toilet

From the one-room rural school to the consolidated school in town

From educational deprivation to the land grant college system

From the ravages of the hookworm to healthy rural children

From mud roads to paved highways

- From self-reliance to a greater degree of centralized decision making
- From uniqueness to mainstream existence
- From the kerosene lamp to electric lights
- From segregation toward integration
- From Rural Free Delivery to the Worldwide Web
- From provincial life to the international dimension
- From male domination to equal rights
- From majority to minority status
- From exploitation of our natural resources toward greater conservation
- From the traditional family farm to the modern business institution
- From a labor-intensive to a capital-intensive operation
- From the image of a straw-chewing rustic to the image of a man in a business suit
- From full-time to part-time farming
- From poverty to virtual income equivalence with nonfarm people

Acknowledgments

Appreciation is expressed to Purdue University; the U.S. Department of Agriculture; J.C. Allen and Sons for photographs; members of the Paarlberg family—Hoey, Don Jr., and Robert—for their counsel; Mary T. Rice, of Purdue's Department of Agricultural Economics, for skillful editing and word-processing; and an unknown reviewer who tipped the scales in favor of publication.

Introduction

If a farmer from Old Testament times could have visited an American farm in the year 1900, he would have recognized—and had the skill to use—most of the tools he saw: the hoe, the plow, the harrow, the rake. If he were to visit an American farm today, he might think he was on a different planet.

The changes that occurred in American agriculture during the 20th century exceeded in magnitude all the changes that had occurred during the 10,000 years since human beings first converted themselves from hunters and gatherers to herdsmen and cultivators. Questions this book dodges are when the 20th century began and when it ended, and when the 21st century begins. In the views of the authors, these are quibbles that have little to do with the flow of the book.

Consider the extent of these changes: In 1910 it took 70 minutes of labor to earn $1.00 of farm income. By 1980, only four minutes were required. In 1900–1914 it took a farmer, with the available horsepower, 81 minutes to produce a bushel of corn. Currently, with mechanical power, it takes less than two minutes.

The farm population in 1900 was 29 million; by the year 2000 it is estimated that it will decrease to just under 5 million. In 1900 farm people constituted 39 percent of the population; by the year 2000, they will constituted just over 1.5 percent. These changes were painful as well as rewarding. "Creative destruction" was the phrase given to the process by Joseph A. Schumpeter, the Harvard economist. The changes were resisted politically, but they were inevitable once science had broken through the barrier imposed by tradition.

The mission of agriculture is to feed an American population now three and a half times as great as a century ago. In addition, with the help of those who produce input items and those who process, merchandise, and export our food, we feed millions of people overseas, from essentially the same land base. Not only does agriculture feed these increased numbers, it also feeds them better. Nutrition-related diseases, such as scurvy, rickets, goiter, and pellagra, have almost disappeared from the vocabulary. The life span lengthens. Young people exceed their elders in stature. Athletic records are continually broken. These changes are attributable in some degree to better nutrition, in part a result of the Agricultural Revolution.

American consumers spend a diminishing share of their income for food; in the early 1900s, over 25 percent was personal consumption expenditure; in the year 2000, less than 14 percent and that for better food, combined now with more services. The enormous achievement of American agriculture during the 20th century is one of our best-kept secrets.

What caused this enormous change? How and why did agriculture break out of the traditional pattern that had restrained it for so many centuries?

The Agricultural Revolution did not originate in the United States. Nor can it be attributed to any one person. But if one person is to be named, it must be Francis Bacon, of England, who in 1620 published his great work *Novum Organum (New Instrument).* Before Bacon, knowledge was commonly sought by uncritical study of Aristotle or the Scriptures. Bacon's new instrument was observation coupled with experimentation. Aristotle's instrument was reason. The Scriptural instrument was faith. Bacon broke with the past. His system was to set up hypotheses, to test them by experimentation, and to reject those that failed the test. Those that survived became the body of science. Bacon said, "I have taken all knowledge to be my province." He ushered in what came to be called The Great Enlightenment, which produced, in the 19th century, the Industrial Revolution. The 19th century also produced those who laid the foundation for the Agricultural Revolution of the 20th century: McCormick, Pasteur, Liebig, Mendel, Darwin.

Mechanical power is a key to revolutionary change. The Industrial Revolution is usually dated from 1776, when James Watt produced his steam engine. The Agricultural Revolution is sometimes dated from 1892, when John Froelich produced the first successful gasoline-powered tractor.

Why did the Agricultural Revolution lag 100 years behind the Industrial Revolution? Perhaps because of the natural conservatism of farm people. Perhaps because the early farm people lacked a margin of safety between trusted traditional methods and the risk of experimentation. Conventional methods had succeeded in achieving survival; what better proof of their merit?

By the 20th century, the gate of tradition could no longer be held shut. When a gate is open, there is no telling who will come through. Through the gate came scientists like Norman Borlaug of the United States, architect of the Green Revolution, and Yuan Long Ping of China, who developed hybrid rice—and quacks like T.D. Lysenko of the Soviet Union, who claimed that acquired characteristics could be inherited. The gate was opened; a newly venturesome people would risk a Lysenko in the hope of getting a Borlaug.

Whatever the reason for agriculture's laggard start toward modernization, once it had made the venture the change was dramatic. The breakthrough occurred not in *additive* fashion but in *multiplicative* form: changes in tillage methods *times* changes in genetics *times* changes in the efficiencies of scale. The total was equal not to the *sum* of the parts but to the *product* of the parts.

The opening of the gate was the genesis of the Agricultural Revolution of the 20th century, the subject of this book.

Agriculture always has had an irreversible mission: feeding hungry people. This book traces the twists and turns, the errors, and the inspired surges of that journey during the 20th century.

How shall we organize a work on the Agricultural Revolution? One might do so by decades, by persons, by institutions, or by epochs. Instead we organize it by the *scientific disciplines*. There were stages in agricultural development; it was not a steady, continuous flow. The several disciplines, overlapping and mutually reinforcing, each nevertheless had its time of special prominence.

First in importance, if not in time, was mechanization, which had its roots in the late 19th century and gained in importance with each decade. Mechanization was basic to the Agricultural Revolution as it had been to the Industrial Revolution 100 years earlier. America, with abundant land and a scarcity of labor, was well suited to the use of labor-saving machinery. Cyrus McCormick, John Deere, and John Froelich were among the pioneers. The critical place of mechanization in the economic, social, and cultural disciplines was asserted by the radical economist Thorstein Veblen when he wrote, in 1921, *The Engineers and the Price System*.

After mechanical power came chemistry, led by Justus von Liebig, who in 1840 published his masterpiece *Organic Chemistry in*

Its Application to Agriculture and Physiology. His book endeared him to the agricultural community, then suffering from a paucity of knowledge.

Then came biology, emerging from the work of Charles Darwin, who in 1859 published his earth-shaking book *On the Origin of Species by Means of Natural Selection, or The Preservation of Favoured Races in the Struggle for Life.* Gregor Mendel read his paper on hybridization to a bored audience in 1865. Louis Pasteur, the great French biologist, developed his cure for anthrax in 1881. And in 1953 came the work of James Watson and Francis Crick, who discovered deoxyribonucleic acid (DNA), "the secret of life." Biology flowered late in the 20th century.

Improved communication dispersed the new knowledge, slowly at first and then explosively, in the closing decade of the 20th century.

Throughout this history, farmers exercised the role of management, putting together the products of these overlapping and reinforcing disciplines, together with finance, economics, and business management, in an effort to make a profit. The separate but interrelated contributions of these disciplines and their integration in the hands of decision makers, both farm and farm-related, are the substance of this book.

Section I

The Technological Revolution

Chapter 1

19TH CENTURY AGRICULTURE

We have called this book *The Agricultural Revolution of the 20th Century*. But is it really a revolution? Most changes in this world are sufficiently gradual that "evolution" would seem a better term. Few beginnings occur suddenly and arrive in full dress, like Athena from the forehead of Zeus in the Greek myth. But the Agricultural Revolution of the 20th century is so drastically different that the term "revolution" seems justified.

There were forerunners, certainly. The 19th century produced great men and women who laid the foundation for the transformation of agriculture that occurred in the 20th century. There must be a gestation period for an idea, just as there is for a living being. What were the agricultural innovations of the 19th century that laid the foundation for the 20th century Agricultural Revolution?

Eli Whitney developed a working cotton gin in 1793 that separated the cotton lint from the seed. A single cotton gin replaced 50 handworkers and greatly increased the demand for slaves to produce the cotton.

By 1807 Robert Fulton's steamboat was in use on American rivers. In 1825 the Erie Canal was opened, and in 1869 the first transcontinental railroad was completed. All three of these developments improved transportation for farm products domestically and to foreign markets and contributed to increased incomes of farm people. In 1903, at Kitty Hawk, North Carolina, the Wright brothers inaugurated powered flight, a new system of transportation that proved feasible for certain high-value farm products.

In 1859 petroleum was discovered in Pennsylvania; this produced kerosene and created a low-cost, transportable fuel for farm lamps and gasoline for farm tractors.

C. Hoxie of Hudson, New York, developed a threshing machine in 1816 to separate grain from the stalk, a task formerly consuming hours of tedious labor. In 1837 John Deere produced a steel plow that would turn the soil cleanly. Before Deere, plows would build up earth on the plow surface, increasing the drag.

In 1834 Cyrus McCormick patented his reaper, which replaced the arduous task of hand-cutting grain. In 1859 J.S. Fawkes produced his steam-driven plow, weighing 10 tons, which foundered helplessly in the loose dirt of the Illinois prairie.

The horse, which was faster, replaced the ox, which was stronger. The cradle replaced the scythe, and out on the western prairie the frame house replaced the sod shanty.

James Patterson, of upstate New York, came to the Horner farm in California, where in 1853 he operated a huge combine pulled by 22 mules. When the combine was put in gear, the mules bolted and wrecked the machine (Higgins, 1950).

Late in the 19th century came the phenomenon known as The Bonanza Farms of the Red River Valley of the North, with one tract totaling 28,000 acres, vividly described by Hiram Drache (1964). These farms failed, were broken up, and were sold to operators of small farms.

Horace Greeley, the 19th century commentator on all manner of things, whose admonition "Go west, young man" motivated a generation of American youth, expressed the varied needs for power that three score years later would be supplied by the Farmall row-crop tractor:

> What our farmers need is not a steam plow as a specialty, but a locomotive that can travel with facility, not only on common wagon roads but across even freshly plowed fields, without embarrassment, and prove as docile to its manager's touch as an average span of horses. Such a loco-

motive should not cost more than 500 dollars nor weigh more than a tun (*sic*) when laden with fuel and water for a half-hour's steady work. It should be so contrived that it may be hitched in a minute to a plow, a harrow, a wagon or cart, a saw or grist mill, a mower or reaper, a thresher or stalk-cutter, a stump- or rock-puller, and made useful in pumping or draining operations, digging a cellar or laying up a wall, as also in ditching and trenching. We may have to wait some years yet for a servant so dexterous and docile, yet I feel confident that our children will enjoy and appreciate his handiwork (Greeley, 1871).

Thomas Nixon Carver was a noted sociologist at a time when a university department of political economy included economics and sociology. He celebrated and anticipated an agricultural revolution when he wrote, in 1911, "Beginning in 1833 there occurred on American soil during the next 30 years one of the most remarkable agricultural transformations ever known in the history of the world" (Carver, 1911). He was correct in what he saw at that early date, but he would have been amazed at what happened subsequently.

Farmers sometimes lagged behind their own technological advances. Corn rows had been spaced wide, 38 to 40 inches apart, to permit the passage of horses. They remained wide for 40 years after the horses disappeared despite the higher yield, with rows narrowly spaced, as new power machinery permitted. Soybeans were planted with the old corn planter and remained wide until the late 20th century, when rows were narrowed or the field was planted solid, with resulting higher yields. Part of the lag resulted from the failure of implement companies to update their machinery. Part of it resulted from the need for wide rows to permit the passage of horses and to provide space for cultivation and weed control. Herbicides were essential before cultivation could be eliminated.

Mechanization was belatedly adapted for special crops: cherries, blueberries, tomatoes, grapes, onions, and almonds. It had special appeal to the young, a factor in the changeover. It related not only to crops; it was general. Mechanized feeding of cattle, hogs, and poultry came in the late 20th century. Barn cleaning, manure disposal, and milking parlors came in. Many farmers air-conditioned their hog barns before they air-conditioned their houses. The argument was that air-conditioning the hog barn was income-enhancing; air-conditioning the house had no such effect. The force of this argument was lost on the housewives.

As farms grew in acreage, distant units were added, so road travel for tractors and combines was a necessity. Farm gates, originally adapted to the horse and wagon, had to be widened to permit passage of the combine. When confinement feeding caught on, fencing the old fields became obsolete.

THE AGRARIAN CREED

Farm people entered the 20th century with the Agrarian Creed, an age-old heritage from earlier times. These are the articles of the creed (Paarlberg, 1964):

1. Farmers are good citizens, and a high percentage of our population should be on farms.
2. Farming is not only a job but a way of life.
3. Farming should be a family enterprise.
4. The land should be owned by the one who tills it.
5. It is good to make two blades of grass grow where only one grew before.
6. Anyone who wants to farm should be free to do so.
7. Farmers should be their own bosses.
8. As agriculture goes, so goes the nation.

This creed would be on the defensive during the 20th century, not so much attacked as eroded. This book traces the impact on that creed brought by the Agricultural Revolution.

FARM AND HOME

The early farm was a unit, not differentiated as between field and fireside. Farm and Home Week, it was called, when people from the countryside gathered at the state agricultural college.

Perhaps the greatest agricultural change from the 19th to the 20th century was the shift of farming from a way of life to a form of business. This occurred gradually, and was resisted by traditionalists and politicians. Here we present some of the characteristics of rural life in the 19th century.

Physical prowess was highly prized. To split the most rails or to excel in a fight were honored achievements. Skill with a gun was a badge of honor. Horsemanship was esteemed. Indian fighters were the heroes of the frontier. Humor was earthy, frequently ribald, and almost exclusively male. There was hard drinking; liquor was plentiful and cheap. Carry Nation used her hatchet to smash saloons.

The Civil War tore the country apart—North against South, state against state, brother against brother.

Photography came in. From many a parlor wall, the framed picture of grandparents looked down, frequently with stern visage.

Families were large. To illustrate: the senior author's father, born in the 19th century, had nine brothers and sisters. The senior author's mother was one of 12. He, a product of the 20th century, is one of four and the father of two.

Self-sufficiency was dominant. Most of what was used by the farm family was produced on the farm. Food was typically what we would call heavy: meat and potatoes in winter, supplemented by garden produce in summer.

Almost everyone had a church affiliation. Religion was a mix of authority and compassion, the two brought together in various combinations, depending on the persuasion of the pastor. They sang:

> Give me that old-time religion,
> Give me that old-time religion,
> Give me that old-time religion,
> It's good enough for me.
>
> It was good enough for father
> It was good enough for mother
> It was good enough for brother
> And it's good enough for me.

Camp meetings and revivals were social as well as spiritual experiences. Occasionally an outburst of religious fervor would sweep the countryside. Alexander Campbell established the Disciples of Christ in 1830. In 1831 the Millerites predicted the Second Coming for 1843 and, when it did not occur, deferred the event to a later date, when it also failed.

School was "readin', 'ritin', and 'rithmetic, taught to the tune of the hick'ry stick." McGuffey's Readers, first issued in 1836, were the standard texts. These graded books carried not only good English but also inspirational literature, poetry, and heroic stories. Religion-based instruction and singing were admissible; the constitutional question was not raised. The one-room neighborhood school was standard. Teachers might or might not have had some training in "normal school." High school was a rarity for rural students, and college was almost nonexistent.

Within the family, discipline was dominant. When there was transgression, correction was invoked.

There was strict hierarchy, white superior to nonwhite, male to female, the elder to the younger, the wealthy to the poor, and the native to the immigrant. In the rural areas, farmers were perceived as more worthy than nonfarmers, whatever their financial status might be.

The telling of tall tales was a folk art; the greater the exaggeration, the more the merit.

Housewives worked extremely hard. The home garden and the poultry enterprise were women's work, in addition to preparing meals, washing, ironing, sewing, baking, churning, and canning. Childbearing came often; midwives were more common than doctors. Death of the mother at childbirth occurred, and the rate of infant mortality was high.

The farm family was isolated. Roads were unpaved and each spring brought "mud time." Out on the frontier, a woman might go into deep depression from loneliness.

Elocution was highly prized. The Fourth of July brought a large gathering and a spellbinder to recite the Declaration of Independence. A school would have an "entertainment," and the chosen student would recite "Give me liberty or give me death" or "The boy stood on the burning deck."

A farm boy was expected to follow the occupation of his father, and as long as free land was available in the West, this seemed a fair prospect. A farm girl was expected to marry a farm boy and to accept without question her husband's decision as to where, when, and how to live.

In the South, blacks were numerous. But, as opportunity offered, they left the farm and took industrial jobs, many of them in the North. To the blacks, the farm was the institution that had enslaved them or their ancestors. In the North, there were virtually no black farmers.

The fireplace provided light as well as heat. The candle was an improvement, and the kerosene lamp was an enormous advance.

Great use was made of daylight; unlike today, the 19th century lived by the sun. The windmill, for pumping water, was a fixture on many old-time farms.

Doctors were few and had limited skills. They made house calls with horse and buggy, sometimes arriving too late. Home remedies were common. People died of "the milk sick" or some other undiagnosed ailment. The germ theory was new and sanitation almost nonexistent. The "backhouse" was a feature for the farmstead, to be tipped over by mischievous boys on Hallowe'en. Life was short.

The parlor was the honored room, and even most modest homes had a semblance of one. There prestigious guests were received, and departed family members were laid out, awaiting burial. There might be an organ.

Group singing was esteemed, and on a long winter night the family might "go through the book." "Barb'ra Allen" and "On Top of Old Smoky" were favorites. The songs of Stephen Foster were much loved. Civil War songs were sung: "Tenting Tonight on the Old Camp Ground" and "Just Before the Battle, Mother."

Hospitals were few and were avoided as "places where a person went to die." There were orphanages, many poorly supervised. Demented people were sometimes kept at home, either cherished by the family or a source of embarrassment. Elderly members stayed on in the family if possible. Three-generation families were numerous. The "poor farm" was for the truly indigent.

Somehow the farm wife found time for quilting or crocheting. There were corn-husk dolls and primitive painting. Self-trained fiddlers played "Turkey in the Straw" at barn dances, after the last year's hay had been fed and before the new hay had been brought in. There was little reading beyond the Bible. Mark Twain was esteemed by the few who read. Poems, seldom read, were the romantic classics, John Greenleaf Whittier and Henry Wadsworth Longfellow, and the country poet, James Whitcomb Riley, with his Hoosier dialect. The artistic urge found outlet, as it often does, even in the most inhospitable setting.

In the parlor might be a small bookcase, with the Bible, a hymn book, some Stephen Foster songs, a set of poems by James Whitcomb Riley, and a large volume titled *Diseases of the Horse*. Such was the literature toward which the plowman homeward plodded his weary way. I like the parallel to Thomas Gray's "Elegy Written in a Country Churchyard."

The farm people of the 19th century had a firm conviction of self-worth. They were engaged in the most important of occupations—food production, a view confirmed by nonfarmers as well as their peers. Thomas Jefferson had said, "They that labor in the earth are the chosen people of God."

With the agricultural revolution of the 20th century, this homely pattern was changed, permanently, for the better, by all objective measurement. But something of value was lost. The senior author raises one cheer. Maybe two. But not three. The junior author raises three cheers.

Sod house on the Great Plains; hollowed-log watering trough, dug well. (Photo: USDA, Centennial Collection)

Harvesting wheat with a cradle. The woman binds the sheaves, twisting the stalks of wheat like twine. (Photo: J. C. Allen and Son)

Cutting grain with scythes. (Photo: USDA, Centennial Collection)

Picking cotton by hand. (Photo: USDA, Centennial Collection)

1849 Archambault horse-drawn Forty-Niner steam engine. Power is on the belt. (Photo of drawing: USDA, Centennial Collection)

Planting tobacco: two horses, three men. (Photo: USDA, Centennial Collection)

Threshing on the home farm in Lake County, Indiana (1930s), using a Huber steam engine, operated by Russell Paarlberg, brother of senior author. (Photo: Don Paarlberg)

A 28-horse team pulling a combine in Washington State before the advent of the tractor. One man, on a raised platform, controls the team. (Photo: USDA)

Three McCormick grain binders in a Wisconsin wheat field. (Photo: USDA)

This Great Plains farmer tries in vain to drive grasshoppers out of his cornfield with a bushy branch. (Photo: USDA)

Threshing with steam power. Coal for fuel (foreground) and the pump wagon for water (background). This is about as authentic as it gets. (Photo: USDA)

Beginning of a dust storm. (Photo: USDA, Centennial Collection)

Eroded land, a discouraged youth, and a substandard house during the Great Depression, Alabama, 1937. (Photo: USDA)

A church in rural New England. (Photo: Robert Paarlberg)

George Washington Carver, born of slave parents in Missouri in 1859, studied at Iowa State College (now Iowa State University) and became known as the "Plant Doctor." Carver developed products from crops such as sweet potatoes, peanuts, and pecans that helped revolutionize agriculture of the South. (Photo: USDA)

Harvey Wiley (in dark jacket), the eccentric Purdue University professor who became chief of the USDA Bureau of Chemistry, organized research that in 1906 led to passage of the Meat Inspection Act and the Federal Food and Drugs Act. (Photo: USDA)

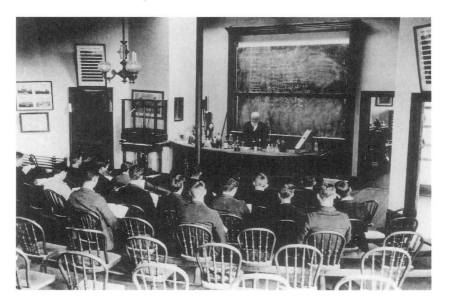

A professor and students of agricultural chemistry in a laboratory lecture, Michigan State University, 1892. (Photo: USDA)

Chapter 2

MECHANIZED AGRICULTURE

Early in the 20th century, at the edge of a village, was a "City Limits" sign, which served as an economic, social, and cultural barrier between town and country. Now that sign is simply a division between two units of local government. Agriculture is losing its uniqueness, a transformation caused by the Agricultural Revolution.

In Chapter 1 we cited the Agrarian Creed, which had been the basic agricultural belief since the founding of the country. How stands that creed after having been impacted by the Agricultural Revolution of the 20th century?

1. *A large proportion of the population should be on farms.*
   During the 20th century, the proportion of the population on farms fell from 39 to 1.8 percent.
2. *Farming should be not only a business but also a way of life.*
   This article of the creed, once the mantra of the farmer, is heard now only in nostalgia.
3. *Farming should be a family enterprise.*
   Farming is still largely a family enterprise, but only if we redefine it to allow for input of nonfarm origin.

4. *The farm should be owned by the one who tills it.*
   By 1996 the per-farm average value of land, buildings, and equipment was $500,000, and much of that was owed to the bank and remained beyond the reach of all but the favored few.

5. *It is good to make two blades of grass grow where one grew before.*
   This idea, fundamental to the motivation of farmers and the purposes of the land grant college system, was challenged by federal farm programs intent on restricting production to offset the price decline caused by the "second blade" and by environmental groups concerned about the new techniques used to produce food.

6. *Anyone who wants to farm should be free to do so.*
   On the face of it, this article of the creed appears to retain its validity. But if a prospective farmer lacks the access to capital and family help, his or her chances of success are virtually nil.

7. *Farmers should be their own bosses.*
   The management revolution now underway brings vertical integration that involves some surrender of individual decision making on the part of the farmer.

8. *As agriculture goes, so goes the nation.*
   Late in the 20th century, agriculture was in trouble while the rest of the country was flourishing.

There you have it. Articles of the old Agrarian Creed have been weakened or wiped out; none of them has been strengthened. They have adapted themselves to the Agricultural Revolution rather than disciplining the revolution, just as some religious creeds have adapted themselves to modern life rather than disciplining the faith of the worshipers.

Nothing has so impacted the old Agrarian Creed as has mechanization, the subject of this chapter. In 1892 John Froelich built the first successfully operating gasoline tractor that powered a Case thresher (Drache, 1964). The gasoline-powered tractor appeared concurrently with the gasoline-powered automobile.

Farmers had given the steam-powered tractor a good try. But it was not well adapted for field work as it was too heavy, too hard to maneuver, and too slow to start; it took a long time to get up steam. A steam engine was costly to operate; it required a tender to supply water and coal. Sparks could set fires in field and farmstead. Next came the ground-driven machine, pulled by horses, with power transmitted to the operating mechanism by a bull wheel.

Early in the 20th century, a number of firms put tractors on the market, names that arouse the memories of older generations: Avery, Allis Chalmers, Big Four, Case, Caterpillar, Cletrac, Co-Op, Dain, John Deere, Emerson, Ford, Ford-Ferguson, Fordson, General Ordinance, Happy Farmer, Hart-Parr, Heider, Holt, Huber, Indiana, International Harvester, McCormick-Deering, Massey-Harris, Minneapolis-Moline, Oliver, Rumely, Samson, Silver King, Titan, Wallis, Waterloo Boy, White, Whitney. During the 20th century, the atrophy of farm tractors was similar to the diminution of the number of farmers.

Most significant of the new tractors was the Farmall, a tricycle-type row-crop machine produced by International Harvester, first sold in 1923, suited for cultivation of corn and cotton. It marked the agricultural transition from horse to machine. In addition, there were various garden tractors plus a variety of homemade tractors converted from cars or trucks.

As with automobiles, many of these early firms are now extinct. In the Midwest, John Deere, Case, New Holland, and Caterpillar are the leading survivors.

Many tractor manufacturers began by making something else; the basic technology of the gasoline engine could be applied easily to many uses. J.I. Case was a threshing-machine manufacturer. Henry Ford made automobiles as well as Fordson tractors. International Harvester, with a full line of farm equipment, had its roots in Cyrus McCormick's reaper.

Soon the bigger tractors came in, such as the 15–30, rated at 15 horsepower on the drawbar and 30 on the belt. Today the John Deere 9400, a huge articulated tractor with power on all four wheels, is rated at 425 horsepower and costs $185,000. The early tractors were often used for belt work, so the horsepower rating reflected that use. But as tractors became more mobile, and as the self-propelled combines came in, belt work diminished and the rating reflected drawbar power only.

The early tractors pulled the machines formerly drawn by horses, such as the grain binder, so that the transition from horse to tractor involved already existing machinery. Over time the tractor and the corn picker were brought together into one self-propelled machine: the combine. And the plant breeder developed grain that would stand stiff and erect so that it could dry on the stem, readied for the self-propelled combine—thus a synthesis of engineering and plant breeding. The tractor pushed the plant breeder to produce cultivars suited to the new age.

Farmers developed affection for their tractors, as they had for their horses. But the tractor's advantage was overwhelming. Tractor power, cited in multiples of horsepower, told the essential story. And with lights and a relay of drivers, the tractor could work night and day. Tractors didn't get tired and or have to be fed during the off-season.

But they were dangerous. On uneven ground, they might tip over, crushing the driver. But horses were dangerous, too. Whether mechanized or traditional, agriculture is a dangerous occupation. In 1993 the National Safety Council rated agriculture the most dangerous occupation, followed by mining and construction.

From 1940 to 1950, the number of tractors increased from 1.6 to 3.4 million (Cochrane, 1979). The number of horses and mules peaked at 26 million in 1919; by 1955 the number had fallen to about 4 million, many of which were riding horses for recreation rather than farm work.

In California during the mid-19th century, combines were pulled by huge teams of horses, as many as 20. This was the horse's last major stand. Mechanization advanced in public awareness. In the early days, agricultural fairs were held, typically county fairs preceding a state fair. Crops and animals competed for prizes. With the advent of the tractor, farm machinery became an important part of the display. The Farm Progress Show, attended by thousands and rotated from state to state in the Midwest, is devoted mostly to farm machinery, with seed and chemicals also prominent.

Tractors were improved as the tasks performed increased and more power was added. By 1930, John Deere introduced the power lift. The power takeoff made entry. By 1938, more than 68 percent of the tractors had pneumatic tires, making for more speed, a smoother ride, better road travel, and, surprisingly, better traction. Horsepower rating rose, allowing tractors to pull heavier loads and further economize on labor relative to land.

Next came numerous small power units, which were electrically driven, and materials-handling equipment, to supplement and make more efficient the larger, more dramatic machines.

The tractor, together with other initiatives, made for an enormous gain in efficiency. Efficiency is measured by total output per unit of total input. Thus calculated, from 1910 to 1994, the efficiency of agriculture increased more than threefold. This advance came at an enormous increase in financial outlay which, however, was more than offset by efficiency gains. In 1939, the per-farm average value of land, buildings, and equipment was $6,000. By 1996, this sum had increased 80-fold. Even accounting for a higher price level, this is an astounding increase.

The tractor appropriately takes the leading role in the Agricultural Revolution. Agricultural engineering departments of the land grant colleges assisted in the development of farm machinery. Mechanization relates to planting, tilling, and harvesting, all of which were transformed, reflecting the increased power and versatility of the tractor. In the corn belt, first there was the single-row, horse-drawn planter; then the two-row planter pulled by a tractor; then the four-row, the 12-row, and most recently, the amazing 24-row planter. Perhaps the ultimate mechanical marvel of the 20th century is the huge, self-propelled, self-leveling combine used on the gently rolling hills of the Palouse region of eastern Washington and Oregon, featured in a 1999 television advertisement of the agribusiness giant Archer Daniels Midland, which portrays itself as the supermarket for the world.

To reduce soil erosion, the idea of minimum tillage appeared, and equipment companies developed machines to do the job. Laser-guided, self-propelled tiling machines laid plastic tile with the desired gradient through an undulating surface. Another use of the laser beam has been to level rice fields to allow uniform depth of irrigation water. Satellites permitted global positioning, and planting equipment was developed that would vary fertilizer placement according to soil needs. Herbicides and insecticides were applied by dusting planes, and rice was seeded from the air.

ELECTRICITY COMES TO THE FARM

Gasoline and similar fuels are but one of several sources of power available to modern farms. Electricity is another. Human labor power was replaced. Electrification of farms proceeded slowly, mostly because of the distance between farms and the cost of installing power lines. The first known rural electric line was constructed in 1906 at Hood River, Oregon. In 1919, only 2 percent of America's farms were electrified. By 1935, only 11 percent were served. In 1936, the Rural Electrification Administration (REA) was created. REA "turned on the lights" in rural America and became the most applauded agency in the U.S. Department of Agriculture. REA was empowered to make subsidized low-interest loans to cover the full cost of constructing power lines and other facilities to supply persons who were without central electrical power. Service was provided on an area basis so that rural industries and rural villages were served, along with farms. By 1960, 97 percent of farms had electricity (Cochrane, 1979).

Electrification reduced labor needs, raised efficiency, and improved the well-being of rural households. The farm operation thus had power for light, milking machines, heating the hog house, ventilating systems, and pumping water. The farm household had power for electric lights, refrigeration, cooking, heating, television, and computer.

There was a 60-year running battle as to whether the REA subsidy for use of electricity should continue or the users should pay full cost. As this is written, the subsidy continues, though at a reduced rate.

Tractors made possible, indeed made necessary, the enlargement of farms and the crowding out of many small farms. "Get big or get out," said the extension man—counsel that was sound but painful. Some farms got big. Some tried to get big and went broke. Some got out and found alternative sources of income. Some small farms stayed in and got poor. In many a country kitchen, the family sat around the table, debating how to cope.

A Testimonial

The Paarlbergs had been farmers in the Netherlands as long as anybody could remember. The senior author's grandfather (great-grandfather of the junior author) was a farmer. In 1847, he was a young boy when his family came to the United States, attracted by the desire to farm and to own farmland. This progenitor had seven sons, all of whom became farmers and none of whom ever seriously considered any other occupation. Each started with next to nothing and became the owner of a farm, small but his own, and provider of a respectable living. These seven sons sired 14 male offspring, only five of whom became farmers. The next generation showed an even greater rate of attrition.

The senior author's home farm of 80 Indiana acres, purchased in 1926, was considered by his father to be enough for himself and his four growing sons. It wasn't. It became 160 acres, then 320 acres, taking over two farmsteads as it grew. The original 80-acre base has now been vastly expanded and is widely scattered, mostly rented. The father's aspiration had been that all the boys would farm; but they left, one at a time. They were unable, or thought they were unable, to amass the capital needed for a farm large enough to be efficient, and they were unwilling to settle for the restricted income from a farm too small to yield an acceptable level of living. There can hardly be a greater commitment to agriculture than this Netherlandish family originally had, and there can scarcely be greater evidence of the impact of the Agricultural Revolution than their leaving the farm. These farm leavers were part of the 33 million people who, according to T.W. Schultz (1978), left farms from 1930 to 1974. "Tractored off" was the cogent phrase. As said earlier, during the 20th century the farm population decreased from 29 to 5 million while the United States population grew from 76 to 270 million.

Given the agricultural revolution, the industrialization of American farms, and the growth of nonfarm opportunities, a reduction in the farm population was inevitable. The reduction was achieved voluntarily by individual choice, based on differential incomes between farm and nonfarm occupations. It could not have been achieved by government fiat. Imagine a government order cutting the farm population and prescribing who could farm and who could not!

The adjustment was particularly difficult for young people in those parts of the country where off-farm jobs were few and educational opportunities limited. In the South, the introduction of the mechanical cotton picker during the late 1950s cut work hours per acre from an average of 66 during 1955–57 to an average of 50 for 1960–61, this in just a few years' time. In addition, herbicides replaced hand-hoeing. The displaced people, mostly African-Americans lacking employable skills, thronged to the ghettos of New York, Washington, Detroit, and Chicago, where many remained unassimilated and unemployed. Nevertheless, a large proportion of all migrants were eventually better off.

Social planners foresaw this problem and knew the solution: encourage industry to come into the troubled rural areas; provide good education for the young. But labor unions, strong in the cities of the North, resisted such solutions, fearing "runaway plants" and the migration of jobs to nonunion areas of the country. Big farmers, who control the farm policy agenda, wanted services for themselves rather than for the disadvantaged, and they wanted an abundant supply of cheap labor. The deprived classes lacked political power. So rural development plans were undercut.

In the former Soviet Union, with its dictatorial powers, the Agricultural Revolution was also taking place, and the number of needed farm workers diminished. This problem was addressed by building nonfarm industrial units into the communes, thus minimizing migration to the cities. This was only a partial and not an economical or socially efficient solution. Many unneeded workers were retained on farms, making for high labor costs per unit of product, both agricultural and industrial. The same thing occurred in China and Eastern Europe.

While it is proper to extol the achievements of the Agricultural Revolution of the 20th century, we should not be blind to what it cost, just as in lauding the Industrial Revolution of the 19th century we should be aware of the pain imposed on the people who had lived by cottage industry. Oliver Goldsmith's poem, "The Deserted Village," written in 1770, paints a tender picture of a happy farm village before commercial considerations destroyed it:

> Ill fares the land, to hastening ills a prey;
> When wealth accumulates, and men decay;
> Princes and lords may flourish or may fade;
> A breath can make them, as a breath has made.
> But a bold peasantry, their country's pride,
> When once destroyed, can never be supplied.

Is there a limit to the number of acres that can be brought together economically in a farm unit? Theoretically, there should not be a limit, but practically, there seems to be. The Bonanza Farms of the Red River of the North did not survive (Drache, 1964). The huge Campbell wheat farms of Montana folded. Large-scale, mechanized, corporate farming operations have failed: Black Watch, Gates Rubber, CBK Agronomics, Multiphonics, and the Great Western Land Company. The enormous British groundnut scheme in Africa failed. The huge state farms of the former Soviet Union ran into trouble. In Brazil, the large-scale Altamira settlement was reported to be in difficulty when the senior author visited there in 1982. The enormous Ludwig pulpwood project on the Jari River had failed, and a collection of Brazilian companies had taken over.

But in former East Germany and in Hungary, mega-farms are surviving as cooperatives. For such farms, the trade-off is between economies of scale in production technology and diseconomies in management.

Farming is a biological business; the conventional wisdom is that farming is unique and unfitted for industrial concepts. There may be limits to the application of mechanized industrialized concepts to a biological undertaking. The poultry business and now the hog enterprise seem intent on challenging the conventional wisdom. The answer to the question apparently must be deferred to the 21st century.

CAN THE FAMILY FARM SURVIVE?

Can the family farm survive the mechanical revolution? Perhaps. But only by redefining itself. Formerly, in the idealized concept of the family farm, the farmer, and his family supplied all the agents of production—land, labor, capital, and management. With the huge farms and enormous investment of the present, it is seldom possible for one person to supply all the factors of production.

But the family farm may survive if we redefine it in practical terms so as to allow for renting land, borrowing money, earning off-farm income, hiring labor, and contracting with integrating firms, leaving the balance of the decision-making role to the farmer operator. The family farm that preceded the mechanical revolution is, to all intents and purposes, dead. Redefined, it may survive.

Chapter 3

CHEMICAL ADVANCES

Chemical use in agriculture has increased dramatically. No one bears more responsibility for the use of chemicals than Justus von Liebig, born in Darmstadt, Germany, in 1803. At that time crop yields in Europe had stabilized at a low level, in a three-field rotation in which one third of the land each season was fallow. This system was hardly changed from medieval times. In the eastern United States, then the only part of the country involved in major agricultural production, early soil fertility was drawn down and yields were declining.

The hunger for understanding about how to restore and replenish soil fertility was insatiable. The processes by which plants received their nutrients were poorly understood. One theory (Jethro Tull's view) was that plants ingested tiny grains of the soil minerals, much as animals were observed to take in their food. A rival concept (the humus theory) held that soil minerals had only a minor role in plant growth; the soil simply supported the plant so that the roots could absorb their true food, dissolved organic matter. Liebig's theory, confirmed by experiment, was different from the other two. He contended that the soil contained minerals that could sustain plant growth and that the humus had to be dissolved into its chemical elements before it could be assimilated. Chemicals could be applied to the soil and could be used directly by the plant, said Liebig. Keen of mind and sharp of tongue, he asked of the advocates of the humus theory: "Humus comes from previously existing plant growth. If a plant must have humus in order to grow, where did the first plant get its humus?" The dispute between those who advocate the humus theory of plant nutrition (chiefly consumers) and those who support the mineral theory (chiefly agronomists) continues to the present day.

There were other disputes about plant nutrition. How much of plant growth comes from the air? What is the role of sunlight in plant growth? Is water a nutrient or simply a conveyance for nutrients? It is astonishing to a modern student of agriculture to learn how recently we have learned about these things.

And animals—why did they thrive on certain rations and languish on others? How did they divide the potential of ingested feeds between growth on the one hand and energy on the other? What was the chemistry of the digestive process? Agriculture could make little progress until there were answers to these basic questions.

Eclipsing these straightforward questions was a deeper issue. Did plants have a vital force, going beyond the province of science? If plants had a vital force, then the study of chemistry would have to be shared with the study of metaphysics. If there were no life force, then questions related to plant growth were wholly within the realm of physical science—specifically chemistry, physics, and biology—responsive to the methods of study prescribed by Bacon in 1620. This issue continued well into the 19th century, when the consensus arose that questions related to plant growth were in the realm of physical science. But the issue is not wholly closed. The gardener who talks to plants and considers them responsive to human moods is a vitalist at heart.

Prior to von Liebig's day, chemists had paid little attention to agriculture, being preoccupied with alchemy and the effort to transform base metals into gold. Liebig brought chemistry to the assistance of agriculture. He wrote six books, the first one a classic: *Organic Chemistry in Its Applications to Agriculture and Physiology* (1840).

The critical chemicals needed for plant growth are nitrogen, phosphate, and potash; a number of minor elements are also needed. Liebig believed that phosphate was the most important and considered both bones and mineral deposits as sources. He discussed the practice of adding sulfuric acid to bones to make the phosphate more readily available, producing what came to be known as superphosphate. He emphasized the importance of potash as fertilizer. He provided the chemical explanation as to why the guano, lime, marl, and gypsum used by farmers were helpful in increasing crop yields. His work helped in the ascendancy of the chemists and physiologists over the vitalists.

But on nitrogen, he missed it. He contended that rain and snow brought down sufficient nitrogen from the air so that there was no need to apply it to the soil. Anyone who has seen the grass green up in early spring after a rain or a snowstorm can well understand that the atmosphere is a source of nitrogen. But that source is insufficient. So powerful was Liebig's misperception about nitrogen that agronomists underestimated its potency for decades.

Here the senior author injects a bit of pertinent history. George Scarseth, a radical agronomist at Purdue University, doubted Liebig's edict about the futility of applying nitrogen. One spring, on the front lawn of the Purdue Experimental Station, where the dean could readily see it, he spelled out (using nitrogen) the word "nitrogen" in letters six inches wide and 10 feet tall. The grass on the nitrogen-treated soil jumped out from the surrounding lawn, dense, tall, and deep green. The message was successfully conveyed.

The chemical industry grew in the late 19th century, bringing down prices of chemicals. Synthetic nitrogen was developed during World War I. There was a rapid increase in the use of nitrogen, as well as phosphate and potash, especially after World War II (Table 3-1). The application of fertilizer skyrocketed, contributing mightily to the spurt in crop production of that time.

Fertilizers are a family of various chemicals combined in a number of mixtures to meet the needs of particular plants on particular soils. They can be spread on the surface, knifed into the soil, sprayed on the plants, drilled with the seed, or side-dressed with the growing crop. They are very potent and diverse.

The increase in crop yields during the 20th century, largely but not exclusively the result of fertilizer use, was prodigious (Table 3-2).

Some farmers lost sight of an economic principle known as the law of diminishing returns. They seemed to think that when some fertilizer was good, more was always better. They applied excessive amounts of fertilizer, particularly nitrogen. When the rate of

TABLE 3-1. Use of chemical fertilizers in the United States.

Year	Nitrogen (N)	Phosphate (P_2O_5) (1,000 tons)	Potash (K_2O)
1930	378	794	354
1935	312	598	307
1940	419	912	435
1945	701	1,556	807
1950	1,237	2,110	1,381
1955	1,933	2,247	1,875
1960	3,024	2,617	2,179
1965	4,638	3,574	2,834
1970	7,459	4,576	4,040
1975	8,600	4,507	4,454
1980	11,406	5,433	6,246
1985[a]	11,502	4,642	5,511

[a]Last year of the series. Fertilizer-use statistics were discontinued. Source: USDA *Agricultural Statistics*, 1987, p. 416.

TABLE 3-2. Multiples of crop yields per acre in the United States, 1900–09 to 1990–97.

Crop	Multiples of Yield per Acre
Corn	4
Wheat	3
Sorghum (or grain)	5[a]
Rice	4
Soybeans for beans	2[a]
Lint cotton	4
Peanuts	3[b]
Potatoes	6
Processing tomatoes	7[c]

[a]1930–39 to 1990–97
[b]1910–19 to 1990–97
[c]1920–29 to 1990–97
Source: Data supplied by G.F. Warren Jr., 1998, from figures obtained through USDA Agricultural Statistical Service.

application was reduced, the yield per acre was maintained as before, and sometimes was increased. During recent years, the use of nitrogen has stabilized while the application of potash and phosphate has been reduced, lessening the degree of groundwater pollution.

Animal Agriculture

Productivity in animal agriculture has been most dramatic in saving labor time. From 1910–14 to 1981–86, work hours per hundredweight of product have been reduced phenomenally (Table 3-3).

For broilers, from 1941 to 1986, feed consumption per pound of product has been reduced by half.

These great gains in the productivity of animal agriculture have come from some unidentifiable combination of better nutrition, pharmaceuticals, reduced death rates, better breeding, better sanitation, better housing, and automation in handling of materials—all results of the Agricultural Revolution.

Killer Chemicals

Another family of chemicals, related to their fertilizer cousins, are the killers. These chemicals kill undesired plants and animals—and human beings—if used carelessly. They are the pesticides: insecticides, herbicides, rodenticides, and fungicides; all are poisons.

Left undisturbed, nature balances the various species of plants and animals with but one objective: the maximization of life. To protect their crops and animals, people attempt to influence nature's balance in their own interest. Despite people's best efforts, these enemies cut food availability worldwide by about 30 percent.

The killers are newcomers in the farmer's arsenal in the mission to meet food needs. First to be considered are the antibiotics, which straddle the border between chemistry and biology. They are the white-hat agents of microbial warfare. The number of microorganisms hosted by the average person is so great as to be almost

TABLE 3-3. Work hours per hundredweight of product.

Years	Milk Cows	Other Cattle	Hogs	Farm Chickens	Broilers[a]	Laying Hens[b]
1910–14	3.18	4.6	3.6	9.5	8.5	2.0
1981–86	0.2	0.9	0.3	2.5	0.1	0.2

[a]1935–39 to 1982–86
[b]Work hours per 100 eggs
Source: USDA, various issues of *Agricultural Statistics*.

incomprehensible. There are from 100,000 to one million living organisms on a square centimeter of human skin. The alimentary canal contains 10 to 100 million bacteria for each gram of tissue. The final census of the living organisms that cohabit with the average person comes to approximately 100 trillion. They make up about a solid pound of living organisms. Under favorable conditions, they are capable of doubling their numbers every 20 minutes. These tiny beings exist in proportionate numbers and in similar kinds in farm animals, such as hogs, cattle, sheep, and poultry.

Some of these germs live in warfare with their host. Some live in combat with one another. Some are symbiotic, either with their host or their fellow microbes. The effects and side effects of these relationships on the human being may be good or bad as we see it; in any case, they are incidental to the microbial purpose, which is to survive and multiply. In a healthy person or farm animal, these organisms exist in a sort of standoff. When there is an imbalance, the person or the animal becomes ill.

In September 1928, Alexander Fleming, a Scottish biologist working in St. Mary's Hospital in London, accidentally discovered penicillin. A spore from the smoky London skies got into his petri dish when he was working with an antiseptic (mercuric chloride) in an effort to kill germs that caused pimples, boils, and carbuncles. To his amazement, the spore from the London smog, not the antiseptic, was killing the germs in his culture. He followed up his chance discovery and named the new killing agent *penicillin*. Research showed that penicillin has the power to attack and kill the organisms that cause strep throat, pneumonia, tonsillitis, scarlet fever, erysipelas, wound infections, gonorrhea, syphilis, anthrax, and gangrene. It was used in treating the wounded during World War II. It is used by veterinarians to treat farm animals.

Honors were heaped on Fleming. In 1945 he was awarded the Nobel prize for medicine. In 1994 he was knighted.

Selman Waksman, a Russian-born microbiologist at the New Jersey Agricultural Experiment Station, discovered streptomycin in 1943. He and his colleagues discovered, in a clod of earth extracted from the throat of a sick chicken, a wonderful therapeutic substance produced by a mold. It proved to be as effective as penicillin, attacking some of the same organisms and others as well: tuberculosis, pneumonia, urinary infections, typhoid fever, dysentery, undulant fever, and wound infections. In 1952 Waksman won the Nobel prize for medicine, as had Fleming before him.

The agricultural experiment stations, impressed with the achievement of one of their own, increased their investigations in microbiology, testing known antibiotics and making applications to plants and animals. Mastitis, an infection in the udders of dairy cattle, was overcome by streptomycin. Bovine tuberculosis, formerly widespread, was nearly eradicated.

Surprisingly, farm animals to which the antibiotics were administered not only threw off their diseases but grew faster as well. So, instead of being limited to therapeutic use, antibiotics were added to the feed of all animals in the herd, well or ill. There were some surprises. The drugs were more effective for young animals than for old, proportionately better for animals on poor than on good rations, more helpful to animals carrying some evidence of disease than to healthy herds, better for animals born as runts than for normal offspring, more advantageous for animals under stress than for those in adjustment, and more rewarding for animals under poor than under good management. Animals raised in aseptic conditions showed no response to antibiotics. The inference was that most farm animals carried some degree of subclinical infection and therefore were helped by antibiotics.

Contagious disease among farm animals had historically been a factor limiting the numbers that could profitably be kept in a single flock or herd. Such restrictions on the size of farm units resulted in high production costs per unit of product since labor, machinery, and building costs were spread across a small number of animals. With antibiotics, this discipline was relaxed. The size of poultry, swine, dairy, and cattle-feeding establishments rose dramatically, and per-unit production costs fell. Protests occurred. Some people resented the transformation of agriculture resulting in part from antibiotics. Animal-welfare people expressed indignation at the poor husbandry made tolerable by antibiotics. Concern arose that the general feeding of antibiotics would cause resistance to develop in the enemy organisms. These newly resistant organisms, it was feared, might escape to the human population, where they would be immune to standard antibiotics.

INSECTICIDES

Another category of chemicals among these battlers in behalf of farmers is insecticides. In 1935, Paul Müller, a chemist employed by the Swiss firm J.R. Geigy, was convinced that a "touch" or "contact" insecticide would be more effective than the oral poisons, such as the arsenates then in use. He discovered dichloro-diphenyl-trichloro-ethane (its name mercifully shortened to DDT).

Müller believed DDT to be relatively safe for human beings. It works through its effect on the nervous system of the organism and is a persistent, long-lived chemical that resists biodegradation. It is amazingly effective.

In past wars, insect-borne diseases took a greater toll than bullets, bombs, and bayonets. Not so in World War II. The Allies learned of DDT and quickly put it to use. Soldiers and sailors by the millions carried small cans of DDT to protect themselves from lice, bedbugs, and mosquitoes. In 1948 Paul Müller was awarded the Nobel prize for medicine.

Farmers, seeing the effect of DDT, quickly expanded its use. From 1939 to 1954, the value of manufacturers' shipments of pesticides increased from $9.2 to $174.6 million.

But as the use of DDT increased, insects began to develop immunity to it. Insects have thus been coping with and overcoming challenges throughout the ages. The rapid insect reproduction rates expedited the development of insecticide-resistant strains. A pair of houseflies reproducing at a maximum rate could multiply to an unbelievable 190 quintillion in just four months. Flies, mosquitoes, and field insects began to develop immunity.

The pesticide companies brought out new chemicals to combat the DDT-immune insects, some more costly and more dangerous than DDT. Soon the market abounded with pesticides. To these new pesticides insects began making their adaptations.

DDT, a persistent chemical, began to permeate the environment. Birds ate insects killed by DDT; the chemical concentrated in their bodies and they died. The whole matter came to a head in 1962 with the publication of Rachel Carson's influential book *Silent Spring*. The doleful title expressed the prospect that DDT and its companion chemicals would kill the wildlife and there would be no birds left. So DDT was banned, effective December 31, 1972.

Crop yields did not decline when DDT was banned. In fact, they continued to increase. Alternative methods of insect suppression were developed: new chemicals, natural controls, and better tillage practices. Integrated pest management, a combination of the above methods and an alternative to "bombing the bugs," was implemented. The predicted adverse effect on agricultural production did not materialize.

Herbicides

Weeds are as old as agriculture. Adam, we are told in the Old Testament, was punished for his sin by having to contend with thorns and thistles. There are about 18,000 weed species in the world that are capable of causing serious economic damage. In 1965, losses from weeds in the United States equaled the combined losses from insects and diseases. In the Third World, more time is spent controlling weeds than on any other farm task.

E.J. Kraus was head of the botany department at the University of Chicago. In 1942 he and his students learned about growth hormones and how they could be manipulated. He repeated the methods used by Fleming, Waksman, and Müller in their discovery of penicillin, streptomycin, and DDT. From the testing came 2,4-D, which was effective in controlling many common broadleaf weeds. Targeted even with small amounts, these weeds became twisted and misshapen, and died within several days. 2,4-D was found to be biodegradable and nontoxic to soil organisms at normal field rates of application. Its half-life in the soil was found to be about two weeks.

The use of 2,4-D expanded rapidly. Whole new families of herbicides were developed; these were increasingly specific for the more troublesome weeds and were tolerated by specific crops. By 1988, some 80 crops were protected by herbicides. Thirty different herbicides are used for corn alone. In 1997 herbicides were used on 96 percent of the corn acreage in the United States, 97 percent of the cotton, 92 percent of the fall potatoes, 97 percent of the soybeans, 90 percent of the spring wheat, and 46 percent of the winter wheat.

The development of herbicides constituted one of the most noteworthy scientific advances in agriculture. One of the great contributions is that with proper use, they permit limited tillage, that is, the growing of crops with minimal disturbance to the soil. On hilly ground, this can reduce erosion by as much as 80 percent. Even on level land, limited tillage checks wind erosion.

Mistrust of Chemicals

In the public mind there is mistrust of chemicals. Some of this mistrust is well-founded; the pollution of groundwater by farm chemicals is a real problem. But some of this alarm is without factual basis, as was the cranberry scare of 1959 and the more recent phobia about the spray residue of Alar on apples. Mightily reinforcing the antichemical mood were the PBB, DBCP, PCB, and Kepone disasters, and especially the calamity in Bhopal, India. The Chernobyl tragedy and the near catastrophe of Three-Mile Island contributed to the antiscience mood.

Two developments were particularly alarming. One was the production and use of 2,4,5-trichlorophenoxy acetic acid, known as 2,4,5-T. When mixed half and half with 2,4-D, this was Agent Orange, used to defoliate forests during the fighting in southeast Asia. Dioxin, sometimes found in 2,4,5-T, is one of the most dangerous of all poisons. It is known through laboratory tests to be teratogenic; that is, it can modify the development of the fetus during gestation, producing abnormal offspring.

Contributing to the antichemical mood was the use of thalidomide taken by mothers during pregnancy, which sometimes resulted in the birth of deformed children. Thalidomide was used in Europe, but its use was forbidden in the United States.

Sarin, the chemical used in the Tokyo subway disaster of 1995, a low-cost weapon of mass destruction, added fuel to the antichemical attitude. Anthrax, also a low-cost weapon of mass destruction (in this case, biological), likewise aroused anxiety. The fusion weapon, the unbelievably powerful agent of devastation equal in destructive power to a million tons of TNT, is capable of annihilating whole areas. It is a product not of chemistry or biology but of physics. These high-tech products caused public apprehension about the uninhibited advance of science.

Regulation of these agents, considered by some to be an invasion of individual rights, is undeniably high in the public interest. An economist would say that the social cost of production is not fully captured in the private cost—a form of market failure.

Organic Foods

The increasing public emphasis on organically produced food is symptomatic of the antichemical mood. There is little medical evidence that organically produced food is superior to food grown with the help of fertilizer and pesticides. Indeed, there is evidence that crops grown on soil supplied with animal manure can carry infections injurious to human health. Organically produced food costs more to grow because yields are much lower. Considering retail prices of 66 different foods, organic foods cost 50 percent more than conventional foods. These findings are confirmed by several authoritative articles (Cooley and Lass, 1998; Duram, 1998).

Farmers, being astute economically, are willing to supply organically produced food if the public is willing to pay a substantial premium for it—corn, 73 percent more; soybeans, more than twice as much (Duram, 1998). In 1994, 4,050 farmers had certified organic operations on approximately 1.13 million acres, less than half of 1 percent of the harvested acreage. Surveys show that by the mid-1990s, the number of acres in organic production of all types had increased steadily. This is no puzzle to agricultural economists; it merely verifies their belief in the effect of incentives.

There is much concern about foods that are alleged to cause cancer. In 1958 the Delaney clause was enacted, requiring the U.S. Food and Drug Administration to require that all chemical additives be proved safe before being used in food. This was the famous zero-tolerance issue; that is, one molecule in an Olympic-sized swimming pool is greater than zero. Despite the extreme nature of this law and the impossibility of administering it, it long remained on the books. Congress members were unwilling to alter it for fear of being punished at the polls by public apprehension about the use of chemicals. One issue here is that as a result of changes in testing methodology, what was zero in 1958 is no longer zero. When the law was passed, laboratory equipment was capable of detecting a suspected chemical when it existed in concentration measurable in parts per million. At present, our ability to identify the suspected chemicals has increased so that we can detect parts per trillion.

Many natural foods, particularly some of the seafoods, little affected by government restrictions, are a far greater threat to health than the foods targeted by law.

There is much public concern about illness from ingesting food. Most food-related illness comes from improper handling of food after it leaves the farmer's gate. Improper cooking, lack of kitchen hygiene, and poor forms of food storage are a more immediate source of illness than the use of chemicals in production.

LOW-INPUT AGRICULTURE

Part of the antichemical mood takes the form of advocating "low-input agriculture." The argument is that a chemically based agriculture, with its intensive use of land, is a threat to the environment.

Low-input (or land-extensive) production is viewed as benign to the environment.

Dennis Avery, of the Hudson Institute, makes a case against this view of low-input agriculture. Without fertilizer and other chemicals, crop yields would be much lower. More acreage would be required to feed the same number of people. Hilly acres and other marginal lands that are highly subject to erosion would have to be used for production. This is happening in many parts of the world, including Africa and southern Asia. Forests are cleared and land is used to the point where it is no longer fit for agriculture. The soil is depleted, eroded, and denuded. With low-input agriculture, the conservation cause would be injured rather than helped, and food prices would be higher. This argument is persuasive to the analyst but seems to have little influence on zealous conservationists. High-input agriculture, with its chemicals, the creature of the Agricultural Revolution, is more a friend than a foe of the conservation cause.

There is no denying the danger of groundwater pollution that results from high-input agriculture or the threat to the resource base that would come from low-input agriculture. The costs and benefits of each type of production must be balanced against one another.

Chapter 4

BIOLOGICAL CHANGES

For the millions of years that preceded the domestication of plants and animals, the governing force that disciplined life was natural selection, which was well defined by Darwin. The fittest survived to produce offspring; the weaker perished.

After domestication of plants and animals occurred, about 10,000 years ago, people began to choose their breeding stock, of both plants and animals. Thus they launched what is called artificial selection, which is based on the well-founded perception that like tends to beget like. By artificial selection, an ear of corn that had been the size of a man's thumb became as big as a coke bottle. Almonds, which are toxic in the wild, became edible. The wild boar became a sleek, domesticated pig. And the wild aurochs, a bovine creature standing seven feet tall at the shoulder, became a tame cow about half as tall—the "foster mother of the human race."

Then, beginning about 100 years ago, came hybridization, started by Gregor Mendel, who in 1865 read his pioneering paper, "Experiments in Plant Hybridization," which was an enormous stride forward in biology. It took a while for Mendel's work to become accepted. Norman Borlaug used Mendelian principles to cross and fix and thus produce his high-yielding varieties of wheat, which contributed to the Green Revolution in Asia during the 1960s. Hybridization of both plants and animals was a major feature of the 20th century's revolution in agriculture.

In 1953 biology took another great stride forward when James Watson and Francis Crick discovered deoxyribonucleic acid (DNA), the life force. The DNA molecule is in the form of a double helix, that is, a double thread coiled like a spring and held together by hydrogen bond cross-links. It looks like a spiral staircase. All DNA molecules have this shape, whether they make up a cactus, a turtle, or a human being. What verification of the oneness of living organisms! The genes, which are the units of heredity, are located on this spiral staircase.

How could they research this mystery, dealing with an entity so small that it could be discerned only with an electron microscope and X-ray crystallography? Watson and Crick worked with theory, integrating the relevant and authentic contributions of biology, chemistry, and physics.

Knowing now the chemistry, structure, and functioning of the hereditary substance, scientists were in position to make applications. The stage was set for modifying the germplasm by techniques variously known as recombinant DNA, gene-splicing, and genetic engineering. During the 1970s, scientists developed techniques for taking genes from higher organisms and splicing them into a piece of bacterial DNA. The resulting new piece of DNA could then be inserted into a gene. By 1980 scientists successfully inserted a new gene into a living animal.

The first applications of the new biology were in medicine. By the late 1970s, researchers had used recombinant DNA techniques to engineer bacteria that produced small amounts of insulin for treating severe diabetes, and interferon, a protein that fights viral infection. By the 1980s, human growth hormones were being produced to treat individuals who would not otherwise reach normal height.

The agricultural people were not far behind the doctors in making applications. Genetically altered bacteria were developed that can inhibit frost damage to farm crops. Scientists have produced mice of greater than normal size. Soybeans have been developed that will survive spraying with Roundup, the herbicide that is deadly to virtually all kinds of weeds. The weed-free soybean fields of the Middle West are testimony to the effectiveness of this new technology. There is a recombinant DNA agent, *Bacillus thuringiensis*, that inhibits the corn borer and checks the spread of the gypsy moth. Genetically enhanced corn was produced on 17 million acres in 1998, 20 percent of the total acreage. About 57 percent of the soybeans planted in the United States in 1999 are genetically enhanced. Dairy farms use biotechnology to produce hormones to increase milk production. This biotechnology-derived agent, bovine somatotropin, is used for almost one third of the dairy cattle in the United States. In Scotland, the sheep Dolly was cloned (that is, made genetically identical to another sheep). If a sheep can be cloned, so can a human being, which raises many ethical questions. Patent laws, intellectual property rights, public underwriting of research, and the privatization of science sped the advance of biotechnology. The speed with which the new biology achieved dominance is shown in Table 4-1.

It is difficult to think of anything more potent for the Agricultural Revolution than the biotechnology that has made its advent in recent years, unless it is the surge in communication, which we consider in the next chapter.

Parallel with DNA and related to it are other products of the new biology that are important to agriculture: embryo transplants, tissue culture, and artificial insemination.

It is clear that the old biology, a single discipline in the tradition of Liebig, Pasteur, and Mendel, has broken out of its mold and has been superseded by something new, which is variously called biotechnology and biochemistry. We now isolate genes and move them about as if they were checkers on a checkerboard.

For farmers, biotechnology lends itself to contractual relationships between farmers and the suppliers of their input items, such as seed for their fields and semen for their animals. And it will connect the farmers in some way with the processors and distributors of their products. Production and sale of crops and livestock of unspecified

TABLE 4-1. Rapidity of the advance of biology.

Dominant Mode of Biology	Years of Dominance
Natural selection	All prehistory
Artificial selection	10,000
Hybridization	70
Biotechnology	?

genetic composition will retreat; contract production, with characteristics specified by the contracting firms, will take their place. This alters the relationship between the farmers and other agents in the production and marketing of farm products. Consumer preferences, which were formerly transmitted through the price system in the competitive market, sometimes failed to get through to the farmers. It took 50 years for dairymen to learn that the consumer did not want milk high in butterfat, and many farmers have not yet received the message. It took far less time, with breeding specific stock specified by integrated firms, for hog farmers to shift from fat hogs to the lean hogs that consumers want.

For the biotech firms, survivors will be the big boys: DeKalb, Dow, Dupont, Lilly, Monsanto, Novartis, and Pioneer or the mergers that will absorb them. The resources needed for competing in the biotech business will be so great that smaller firms are unlikely to survive as independent units, which is one explanation for corporate mergers.

For the agricultural experiment stations, the task will be to provide the big biotech firms with people who are well trained in the basic sciences. This has already been the case for hybrid corn. The role of the publicly funded land grant college, which once focused largely on applied work at the farm level, is shifting toward basic research, to the advantage of private firms. This is raising questions about the interface of private and publicly funded work in agriculture. To what extent should tax dollars be used to finance research at universities that is marketed and sold by private firms? The agricultural experiment stations are shifting toward work in science, nutrition, environmental matters, land use, fish and wild life propagation, rural and international development, forestry, nursery production, and even tourism—all to the bewilderment of their original clientele, the producers of crops and livestock.

Some consumers, especially those in Europe, have become alarmed by "genetically modified organisms" (GMOs). Protests have arisen about importation from the United States of corn and soybeans that have been genetically modified to resist weeds and insects and about importation of meat from animals that have been fed hormones for faster growth. The scientific base for this concern is shaky at best, but its potential for foreign trade is enormous.

Clearly, we are on the threshold of a new epoch in agriculture. The Agricultural Revolution has entered a new phase, the biological era, that is being built on the mechanical and chemical stages.

Animals

We have been discussing, for the greater part, biology as related to plants. Now we turn to animals, including the human animal.

Before Mendel, the continuity of life between parent and offspring was a subject of awe, wonder, mystery, riddle, or perceived doctrine—anything but science. In breeding animals, the pure line was thought to be best. An animal of mixed breeding was a "scrub," a term of derision. It is possible that biblical admonition, then very powerful, was influential in this attitude. The Book of Leviticus says: "Thou shalt not let thy cattle gender with a diverse kind; thou shalt not sow thy field with mingled seed" *(19:19)*.

In the human population, a "half-breed" was an inferior person. The male was thought to be the more important of the two parents, stamping his attributes on the offspring. Tracing the male parentage was for many years the accepted way of recording ancestry for human beings and other animals.

One sees here a reflection of the racism, sexism, and aristocratic pretensions that characterized human institutions throughout the ages. Mendel's work would destroy the basis for these beliefs. He proved that offspring received their attributes equally from the male and female parent. But correction would be a long time coming. Farmers, under pressure to make their operations pay, would slough off these erroneous ideas sooner than the leaders of the social system. A hundred years after Mendel, livestock ranchers had practically abandoned the purebred idea. And 385 years after the *Mayflower,* social elites were still tracing their ancestry to those who landed at Plymouth Rock, though they might have received only a tiny fraction of 1 percent of their biological inheritance from that remote ancestor. Nature, in her wisdom, keeps broadening the genetic base. Human beings, in their ignorance, keep denying the basic fact.

Livestock breeders found that crossbreeding, say of purebred Hereford with purebred Angus cattle, produced offspring with hybrid vigor (heterosis). Mendel's principle of hybridization, which he developed through research on plants, worked also for animals, including (to the consternation of the elites) the human animal. The genetic nature of heterosis is still a mystery.

Livestock breeders learned, just as plant breeders had, that a cross can be made and that the superior nature of that cross can be fixed by careful selection through a number of generations. Corn breeders had thus fixed the character of nutritious high-lysine corn that could be planted year after year without buying new seed corn. Poverty-stricken Third World farmers thus could supply themselves with a nutritious food without the need for cash to buy seed each year when there was no outcrossing. Hog breeders thus fixed the character of lean hogs, to the benefit of themselves and consumers.

Artificial insemination vastly increased animal efficiency. Up to 25,000 cows have been bred artificially to one bull in one year. The benefit of a superior sire could thus be multiplied. The dangers are multiplied as well, as some breeds of swine have learned. Hidden genetic weak traits can be spread throughout a particular breed.

A successful biological initiative, the development of hog cholera serum, was demonstrated at the U.S. Bureau of Animal Industry at Ames, Iowa, in 1907. It is estimated that in 1913, 10 percent of the hogs in the United States died from hog cholera. Today hog cholera is virtually nonexistent; there is careful monitoring lest it break out anew. Vaccination is no longer necessary.

A creature not usually considered an animal is the fish. The Agricultural Revolution, with its knowledge of nutrition, sanitation, engineering, and genetics, has produced aquaculture, the commercial raising of fish. From 1975 to 1994, the raising of domesticated fish increased fivefold, from 130 to 651 million pounds (Kernan, 1997). Fish have several advantages over land animals. They are cold-blooded and need not use energy to keep themselves warm, which cattle and hogs must do. They are waterborne, thus movement requires less effort. Consequently, the rate of feed conversion is more favorable than for land animals. The Agricultural Revolution has produced a new farm enterprise.

With the interaction of chemistry and biology, better animals ingest better feed and are better protected from disease; they produce meat, milk, and eggs that are better suited to the consumer's need and wish. It is a win-win situation.

GERMPLASM

Biotechnology has a metaphysical component. Germplasm is the only element in the universe of living creatures that is eternal. The germplasm borne by a human being is a blend of germplasm carried by his or her progenitors, male and female. In the absence of mutations, this goes back an untold number of generations. In England, genetic testing of a mummified man of 1,500 years ago led to the discovery that his descendants lived within a mile or so of his grave. Are we giving offense to human beings or the spirit in interfering in the transmission of genetic material? The question is fair, and the answer is not immediately obvious. Every form of human interference with the reproductive process, whether artificial selection, hybridization, or genetic engineering, is vulnerable to the charge of genetic manipulation.

Are some areas of scientific investigation off limits? Is there such a thing as forbidden knowledge? To scientists this question is an oxymoron; to them, no area of investigation is off limits. The ancients thought that some things were forbidden. In the Genesis story, Adam and Eve were driven from the Garden of Eden for eating the forbidden fruit of the tree of knowledge of good and evil. Pandora was forbidden to open her box but did so and released pain and suffering into the world. Prometheus stole fire from the gods and was punished horribly. Did the ancients have some insight as to which forms of inquiry were legitimate and which were not? Or have we, in the progress of science, shown the "insight" of the ancients to be an illusion?

Is there some rate of change so great that we are in danger of flying apart? Are we, with excessive rates of change, placing greater stress on our social, economic, cultural, and political institutions than they are able to bear? Does the economic principle of diminishing returns apply to the rate of technological change as it does to the rate of fertilizer application? If biotechnology is good (which it is), is more biotechnology better and is the most biotechnology imaginable the best possible goal? A vague feeling of unease enters the minds of many as the biotechnology revolution accelerates.

The wise old heads counsel that it was always thus when new ground was plowed and that the surge of biotechnology is no different. But is it different?

Questions about the appropriateness of the headlong rush of biotechnology are legitimate, and the answers are not immediately evident.

Chapter 5

INFORMATION EXPLOSION

Like other changes described in this book, the Information Revolution can be traced to several founding fathers. One is Johannes Gutenberg, who in 1453 developed the printing press. After that, printed material became so cheap that even the poor could have it. Reading and learning expanded. Indeed, the modern age is sometimes said to have begun with the printing press. Reading and learning were no longer the exclusive possessions of the wealthy. Information that had formerly been transmitted orally could now be printed and widely distributed.

Another pioneer was Samuel Fairbanks Morse, who in 1844 sent the telegraph message between Baltimore and Washington, "What hath God wrought?" Information could be sent by wire.

Alexander Graham Bell transmitted the spoken word along such wires in 1876, and the telephone came into being. In 1901, Guglielmo Marconi eliminated the wires, and the radio was developed. In 1925, John Logie Baird transmitted recognizable human features—the beginning of television.

In the late 20th century came the information explosion, the computer revolution: the facsimile (fax), the transistor, the microchip, the

Worldwide Web, satellite imagery, and e-mail. The world was awash with information.

Formerly, information had some endorsement before it reached the public. Scientific publications were peer reviewed. Books were edited. The newspaper had a city editor who screened reporters' submissions. But with the Worldwide Web, anything and everything became public, from solid scientific knowledge to private ranting of angry cranks to gossip about a president's affair with a White House intern. Responsibility for screening this vast outpouring shifted from some accountable central point to the intellectual and ethical resources of the individual person.

Before the information explosion, the progress of shared intelligence was gradual, advancing by stages with better roads, the initiation of rural free delivery, the mail-order catalog, the county paper, the Agricultural Extension Service, the chautauqua meeting, the political conventions, and the coffee shop. But with the eruption of the Computer Age, information became so omnipresent and undisciplined as to overwhelm the average person.

Information is a critical factor affecting the economics of agriculture. Firms need to know costs, locate buyers of their products, and be able to obtain inputs. Information is usually costly to obtain and is often costly to process and deliver. For some activities, such as transmission and delivery, per-unit cost falls as the volume increases. That is, there are economies of scale in information. Information often has a location and timing advantage—that of being in the right place at the right time. For goods of similar quality, like many farm products, early access to and action on information creates profit opportunities that otherwise would be lost. These factors were already quite important at the beginning of the 20th century, but over time the Information Revolution has lowered the cost of obtaining, processing, and delivering information while increasing the speed at which that information flows.

Farmers need information about the commodity markets, performance and prices of the machinery they consider buying, the politicians they elect, plant and animal nutrition, the size of the crop in the United States and other countries, and the weather. They need the best guess of the outlook person who assesses the prospect for these things. Farmers rely more on information than do people in almost any other vocation. And much of what they need is readily available on their computer.

The Information Revolution lowered the cost of obtaining, processing, and delivering information while increasing the speed with which it flows. At the beginning of the century, when information was costly to obtain and location was important, certain places developed as trading centers for various agricultural commodities produced in the United States. Chicago was the center of grain and livestock trading. New York specialized in imported commodities such as sugar, coffee, and cocoa. Minneapolis and Kansas City focused on hard wheats.

These trading centers had characteristics that facilitated the flow of information. They provided a central point where large numbers of both buyers and sellers could meet. That raised the volume of transactions and generated more information about market conditions. All had strong connections to the rest of the United States and the world beyond. Those connections involved transportation, by road, rail, water, and air. Other connections were via print media and telegraph-telephone service. Each trading center was not far from the primary source of supply. This improved knowledge of growing conditions, transportation problems, and storage capacity.

Location was critical because the collection and delivery of information was largely a private affair. Though the United States government collected and published some statistical data, coverage was limited and publication infrequent. The days of massive government collection of statistics had not yet arrived. It paid to be with other traders near to rural producers yet connected to the world beyond.

Decentralization of Markets

The Information Revolution has reduced the advantages of a centralized location. Consider Chicago and the U.S. meat industry. At the start of the century, had one asked what city was the capital of the U.S. meat industry, the answer was clear—Chicago. Chicago had the rail terminals—the largest stockyard for collecting animals. It had the traders. It had the meat processors, such as Swift and Armour. Other markets, such as Omaha, St. Louis, Indianapolis, and Cincinnati, were the places where animals from local markets were pooled for shipment to Chicago.

Futures trading for livestock is still centered in Chicago. The stockyards are gone, having been replaced on a smaller scale by stockyards in Omaha and other cities farther west. Meat processors left Chicago and are now dispersed throughout the rural countryside.

While these changes occurred for a variety of reasons, it is the Information Revolution that allows this system to operate. Hog buyers at a processing plant in Logansport, Indiana, know instantly the prices for live hogs at any of a number of markets around the United States. One glance at the computer screen can tell them what they need to know about how prices differ from market to market. If there is a sudden transportation problem, that information, too, will appear and they can telephone to arrange an alternate supply. If it is unusually hot in the Middle West today, buyers will know where and can assess the impacts of weather on hog shipments. The person selling a certain processor's pork sees this same information. So do buyers and sellers at plants and markets around the United States and the world. Electronic contact has replaced physical contact.

The decentralization of a number of agricultural industries, like that of meat described above, happened because the Information Revolution reduced the costs of obtaining, processing, and delivering information so that other forces determine location. This was an evolutionary process, which continues. Some have argued that centralized trading centers like the Chicago Board of Trade will wither as the Information Revolution continues.

There appear to be two main phases in this revolution. The first phase begins around the time of World War II and marks the beginning of the decentralization of processing activities. This phase corresponds to the arrival of the telephone and telex, which allowed buyers and sellers to obtain information without direct contact. Improved national and international communication sped the delivery of information to remote locations.

The second phase recently arrived with the personal computer, fax, e-mail, the microchip, and satellite data transmission. As with the first phase, it is not necessarily the cost of collecting information that has been reduced. In many cases, that remains labor-intensive and costly. Rather, the costs of transmission, delivery, processing, and reproduction are falling.

Grades and standards established by the U.S. Department of Agriculture (USDA), coupled with modern communication facilities, made it possible to buy and sell farm products without seeing them, thus saving an immense amount of time and travel. The USDA provided estimates of planting intentions, growing crops, and storage stocks, as well as livestock numbers, breeding plans, and rates of slaughter. These estimates were distributed to farmers free of charge and were held in high regard by traders in the central markets. The market was made more intelligent and less vulnerable to rumor or manipulation.

At the beginning of the 20th century, information was labor-intensive and costly to obtain for U.S. farmers. Local newspapers may have contained news of the past few days or the past week's commodity prices but sometimes were not delivered to rural households. The primary source of outside information was a visit to town. When farmers were favorably located, with effort they could visit two or more towns and check on prices at different elevators, lumber yards, hardware, or other input suppliers. Marketing a crop consisted of an early morning departure to town with a commodity for sale at the price it would fetch.

Transformation of Marketing Institutions

In the early decades of the 20th century, on the farm where the senior author grew up, there were various ways of marketing products: for tomatoes, a contract; for milk, a price negotiated between the Milk Producers Association and the milk processor; for asparagus, consignment to an agent in the South Water Market in Chicago; for hogs, a commissioned sale at the Chicago Stockyards; for eggs, exchange for groceries at a price arrived at who knows how; and for cabbage, haggling on price with a huckster. None of these methods of sale had electronic input, and none was very efficient. Consolidation of markets, specialized production, better information, electrification, and improved roads helped modernize the process.

Regular mail allowed farmers to track local, national, and international developments. Telephone service allowed farmers to determine prices at nearby elevators without actually visiting the elevators. Improved roads extended the potential range over which a farmer could deliver commodities. Less effort could be devoted to gathering market information, and farmers could select from a wider variety of locations at which to buy or sell inputs.

The spread of radio accelerated this process. In the evening, farmers could follow that day's price and market developments while planning for tomorrow. At noon, farmers could learn of the morning's price changes. The noon farm news detailing prices and volumes moving through various markets in Chicago and elsewhere was an ever-present feature of the farm kitchen in the 1950s. Farmers may now have a two-way radio in their truck or the cab of their tractor or combine and a cellular phone to confer with fellow workers or with prospective buyers or sellers.

The telephone, radio, and (to a lesser extent) television remained the primary tools for gathering off-farm information until the end of the 1980s. At that time communication technology developments outside of agriculture profoundly affected how farmers interacted with the outside world. During the 1960s and 1970s satellite communication allowed the nearly instant transmission of events from around the world directly to the farmer. Droughts in Africa and floods in Brazil were no longer events learned about days or weeks later; they could be reported as they happened. The transistors and tubes that made electronic equipment large and bulky were replaced by miniature chips that reduced size and increased capacity and mobility. Wiring and cables were replaced by fiber optics; computers shrank in size and cost but increased in capacity.

The result of these developments was that large volumes of instant information and data could be transmitted to remote locations at declining cost. Entrepreneurs were quick to offer services whereby the individual farmer could, for a fee, obtain nearly instant market, weather, and news information. In addition, these firms provide farmers with commentary and analysis that were formerly the domain of the broker.

Previously, farmers relied on occasional radio news and telephone calls to local elevators. Now they can check the markets and weather throughout the United States while unloading grain. Using these services, farmers have an unprecedented view of global commodity markets.

Information at the Farm Level

The Information Revolution on the farm was not confined to obtaining off-farm information. Within the farm operation, a revolution occurred as well and spawned a new discipline—agricultural economics. One difficulty facing farms at the beginning of the century was that farmers did not have accurate records of expenses or agronomic relationships. Early work in farm management at land grant colleges focused on developing a system of record keeping. This provided individual farmers with more knowledge about their operations and, when combined with records from a number of farms, provided the data needed to study and improve management techniques.

The advent of the personal computer in the 1980s led to further improvements. The computer can store and provide easy access to much more information. Calculations of relationships could be performed quickly and accurately. The computer's capability allowed farmers to consider more farm-planning options. Still, someone had to spend the time to put accurate data into the computer. Collecting the information remained labor-intensive, while processing and delivery costs fell.

A recent development allows a public or private central service to receive price and quantity input from a farmer for various items of sale and purchase and summarize and analyze the data for profit or loss in comparison with other farms.

Another technology making its advent as the 20th century closes is the Global Positioning System (GPS). This system uses satellite tracking and onboard computer monitoring to control various field operations. Sensors can adjust fertilizer application to soil needs of specific areas of the field.

Information is the raw material on which farm operators rely in their effort to make their farms pay. The information explosion is a key dynamic of the Agricultural Revolution.

Chapter 6

MANAGEMENT: INTEGRATION OF THE DISCIPLINES

Scientific disciplines set forth in the preceding chapters must be integrated with one another, farm by farm. Integration is the task of management. Science remains in the laboratory unless there is the incentive to adopt the knowledge. This is the difference between science and technology. Economics is the integrator.

Like the other disciplines cited as central to the beginnings of the Agricultural Revolution, economics had its founding fathers. Adam Smith, the Scottish savant, wrote *The Wealth of Nations* in 1776. There also were David Ricardo, John Stuart Mill, W. Stanley Jevons, Leon Walras, and Alfred Marshall, the English professor who wrote *Principles of Economics* in 1890. Another Englishman, John Maynard Keynes, published his *General Theory of Employment, Interest, and Money* in 1936. This book helped the United States emerge from the Great Depression.

Economics, thought by some to be a jumble of jargon, is in fact a well-ordered body of principles, which may be summarized as follows (Paarlberg, 1968):

- *Key Economic Principles*
- *The Principle of Incentives*
  Other things equal, people will gravitate to activities whose financial rewards are greatest.
- *The Law of Demand*
  As price rises, the quantity demanded decreases.
- *The Law of Supply*
  As price rises, the quantity supplied increases.
- *The Interaction of Demand and Supply Determining Price*
  Price is discovered at the point where demand and supply are equal.
- *Price and the Cost of Production*
  In the long run, the price equals the cost of production, not farm-by-farm or year-by-year. During the terrible '30s, the cost of production was greater than the price. During the euphoria of the '70s, the price was greater than the cost of production.
- *The Law of One Price*
  If markets are open and after transportation and other transaction costs are allowed for, only one price will prevail in a given market.
- *The Principle of Diminishing Utility*
  The satisfaction gained from use of the second of a series of items is less than the satisfaction obtained from use of the first.
- *The Law of Diminishing Returns*
  If the input of one resource is increased by equal increments per unit of time while the inputs of other resources are held constant, the total output will increase, but beyond some point the resulting output increases will become smaller and smaller, and may eventually become negative.
- *The Principle of Opportunity Costs*
  The cost of producing one article is the sacrifice of the opportunity to produce another.
- *The Law of Comparative Advantage*
  This is best illustrated by an example. Indiana produces more wheat and more corn per acre than does Oklahoma and so has an *absolute* advantage in the production of both crops. In Oklahoma, wheat yields are 85 percent as much as corn. In Indiana, wheat yields are only 37 percent as much as corn. Thus, wheat

does better relative to corn in Oklahoma than in Indiana. So Oklahoma specializes in wheat, while Indiana specializes in corn. Oklahoma does not have an *absolute* advantage in the production of wheat but has a *comparative* advantage.
- *Water and Molasses*
  Water finds it level quickly. Molasses finds its level, but slowly. Economic theory generally states adjustments as if they found their level quickly, as water does. Actually, the adjustments come slowly, as with molasses.

These principles have come under heavy attack for assuming that other things are equal when they are not, and for theorizing that the desire for profit is the only motivating force. These criticisms are certainly valid. But the principles, despite their shortcomings, explain much of what goes on in American agriculture. They help bring together the elements of the Agricultural Revolution that were described in earlier chapters.

Subfields of economics emerged: labor economics, industrial economics, and agricultural economics. The latter, called "farm management," first appeared at about the beginning of the 20th century, led by G.F. Warren of Cornell University, H.C. Taylor of the University of Wisconsin, and George Pond of the University of Minnesota.

The early farm-management workers greatly expedited the transition of agriculture from "a way of life" to "a business." They developed farm accounting after the manner of business accounting. They made a charge for the use of capital even if all the capital were owned by the farm operator. They made a charge for farm labor even if the farm labor (wife and children) received no pay. They tried, but did not succeed, in separating the farm operator's return for labor from his return for management. They reported "labor income," which was the combined return for labor and management. They did cost accounting, with debits and credits, attempting to determine the return to individual farm enterprises.

How much they contributed to the demise of the idea that agriculture was a way of life is uncertain, but it was substantial.

As agriculture entered the mainstream of economic life, farm management became "agricultural economics" and then "applied economics" or "resource economics." It is to this evolving discipline that both the senior and the junior authors belong, their fields of vision traceable to the pioneering work of Adam Smith and Alfred Marshall.

It is good to realize that changes occur not in random fashion but in accordance with a set of known laws. What may seem chaotic is often the working out of poorly understood principles.

Operating with comparative advantage, the Midwest came to produce feed grain and livestock. The Great Plains chose wheat; the South, cotton; and Wisconsin, dairy. California, with a great variety of soils, water, and climate, chose to produce 200 different items. Comparative advantage was decisive in determining land use.

In accordance with these incentives and disciplined by profit and loss, farm prices and costs were determined by the laws of supply and demand, large farms replaced small farms, the off-farm migration occurred, the tractor replaced the horse, and agribusiness took over some of the decisions previously made by farm operators. Production of soybeans expanded, production of oats diminished, and Holstein cows replaced Jersey cattle. Government policies—such as the Homestead Act, price supports, and production controls—modified the basic pattern.

When input prices rose, technological change reduced their effect. When diminished returns threatened, technology often was able to cope. Scarce labor was relieved by machinery, scarce land by fertilizer. It was to the individual farmer's benefit to adopt new cost-reducing technology. And as technology advanced, supply grew faster than demand, and the price fell. Thus, farmers found themselves on a technology treadmill, trying to adopt technology as quickly as possible to stay ahead of the falling prices.

Each farm is in a sense unique, with its particular endowment and peculiarity, so the economic principles enunciated previously do not have universality. The individual farmer tries out various practices, sometimes succeeding and sometimes failing. A farm is, in effect, an experiment station of which the farmer is the director.

Crop yields per acre increased phenomenally, as indicated in Chapter 3 (Table 6-1). Is it possible to identify the separate effects of the various disciplines in bringing about the increased yields of the 20th century? G.F. Warren Jr., retired professor of horticulture at Purdue University, endeavored to do so with the help of 50 crop specialists, a most difficult undertaking because of the interrelationship and the multiplicative nature of the relationships. Nevertheless, here is a very rough attempt on the part of the authors, based on Warren's work (Table 6-2).

The divisions between the respective disciplines are arbitrary. For example, agricultural chemistry has become biochemistry, making it difficult to know what is chemistry and what is biology.

TABLE 6-1. United States average yield per acre for 10-year periods and the last eight-year period.

Period	Corn for Grain (bu)	Wheat for Grain (bu)	Sorghum for Grain (bu)	Rice (lb)	Soybeans for Beans (bu)	Lint Cotton (lb)	Peanut (lb)	Potato (cwt)	Processing Tomatoes (ton)
1900–09	27	14.4	NA	1,500	NA	185	NA	57	NA
1910–19	26	14.1	NA	1,680	NA	184	770	58	NA
1920–29	27	14.0	NA	1,850	NA	163	708	67	4.5
1930–39	24	13.3	12.7	2,180	16.1	206	708	68	4.2
1940–49	34	17.1	17.5	2,080	18.9	266	704	101	5.9
1950–59	44	19.7	23.8	2,810	21.4	362	976	165	10.7
1960–69	71	26.4	47.8	4,050	24.9	478	1,520	204	16.4
1970–79	89	31.4	54.0	4,550	27.9	475	2,380	248	21.6
1980–89	106	35.8	59.8	5,140	30.3	577	2,480	284	26.4
1990–97	121	37.6	65.0	5,760	36.5	654	2,390	325	32.3

Source: Calculated by G.F. Warren Jr. as straight averages from USDA National Statistics Service Data. The help of Ralph W. Gann, State Statistician, USDA, Purdue University, and his staff in providing the original data is gratefully acknowledged.

TABLE 6-2. Degree of importance in yield during the 20th century. (number of pluses indicate the degree of effect)

Crop	Multiples of Increased Yield	Mechanical	Chemical	Biological	Management
Corn	4	+	++	+++	+
Wheat	3	+	++	+++	+
Sorghum	5	+	+	++++	+
Rice	4	++	+	+++	+
Soybeans	2	+	++	++	+
Cotton	4	++	++	++	++
Peanuts	3	+	++	+	+
Potatoes	6	+	++	+	++
Tomatoes	7	+++	+	++++	++

Little has been said about irrigation, which has increased yields of corn in Nebraska, cotton in the mid-South, and tomatoes in California, altering the competitive relationships of one area to another and reducing production costs and prices.

In 1934, when the senior author was in his twenties and was farming with his father and brothers in northwestern Indiana, he wrote the following article, which was printed in the *Lake County Star.*

QUALIFICATIONS FOR A FARMER

First of all, a farmer has to like his work. If he doesn't—if he doesn't like to get his hands dirty and if he doesn't like long hours, open fields and solitude—then he'd better let farming alone and try something easy, like being a lawyer or running a nonfarm business.

A farmer has to be an optimist, too, to tide him over those periods when his commonsense would tell him that everything had gone wrong. He has to have a reserve of hope that he can call on to bridge him over at a time like that. You have to measure a farmer by the amount of punishment he can take.

You can make a businessman in five or six years and a doctor in eight or 10, but it takes a lifetime to make a farmer.

A farmer's education begins the day he is old enough to stir the soil and continues until he is laid to rest in it.

A farmer has to be a manager, to choose his crops and plan his schedule in order to save time and eliminate waste motion. He must be able to distinguish between things of prime importance and secondary detail. He has to be able to live within his income and without it when he has none. He has to have a practical working knowledge of a wider range of subjects than any professional man. His is an education in its most practical sense.

A farmer has to know the land intimately, like a living thing—its chemistry, its peculiarities, when to plow and when to plant, how to nurse a tender crop through frost and flood, through drought and heat, how to wrestle a living from the stubborn soil. He must know the weeds and how to fight them, insects and plant diseases, the weather and its effects. He must know all the breeds of livestock, how to feed and care for them, how to treat an animal for common ailments, how to recognize symptoms and diagnose diseases. He makes better use of his knowledge of the principles of breeding and sanitation than the human race does for its own stock.

A farmer has to know the markets, when to buy and when to sell. In normal times, a farmer buys a wider range of goods than anyone else, which means that he has to have a well-developed sense of values. The farmer figures in every branch of economics: he is present in every business deal; he is a factor in every political question. To keep abreast of the times, he must know and understand the working of everything that touches his life. He must know which groups are his friends and which are his enemies—whom to trust and whom to fight, whom to work with and whom to oppose.

A farmer has to know machinery and how to care for it. He has to know how to adjust a plow so it will turn a clean furrow with the least draft. He has to know how to start an engine on a zero morning, and how to patch up a piece of machinery to keep it going and finish the job until he has time to fix it. He must learn all the laws of physics by experience—he must be able to engineer any job from laying a string of tile to leveling up the foundation for a new shed.

A farmer has to be his own boss, his own carpenter, blacksmith, mechanic, electrician, plumber, mason, and painter. He has to be able to hire men, boss them, and use them to good advantage. He must know how to build a new fence or repair a leaky roof, how to set a shock of grain so it'll stand and how to stack hay so it won't rain in. He has to know how to handle horses—how to get the most work out of them without abuse; how to tell when they have reached the limit of their endurance. He must know how to break a colt to make a good horse of him. He can't be a good farmer if he doesn't have a way with horses.

A farmer must know the feeling of frustration that comes from working with poor stock and poor equipment, when he knows so well what he should have. His is the continual philosophy of "doing without."

A farmer has to work with his head and his hands, with his heart and with his soul. So the next time someone tells you he is a farmer, try to look past his dirty overalls and see the whole man.

That article was good enough for its time. But since then the farmer would have had to learn to drive tractors rather than horses, run a combine rather than a grain binder, and use minimum tillage instead of the plow. Many of his valued skills, like husking corn and shocking grain, would have become obsolete. He would have had to learn about pesticides and new kinds of fertilizers. Very likely he would be wearing slacks or even a business suit instead of overalls. His farm might have increased in size and he would have borrowed money to finance the expansion. The long list of his own manual skills would have been replaced by experts: veterinarian, building contractor, farm-implement dealer, and fertilizer consultant. He would have a computerized record keeper. The idea that a farmer would have to serve apprenticeship as a farm youth is no longer valid. Poultry integrators recruit their producers from factory workers and mail clerks who would not have to unlearn the outdated practices they had acquired on farms in their youth. Most probably, the modern farmer's home would have been improved with bathroom, electric lights, and television. He would have had to extract income from the farm, somehow, to make these improvements. Each of these changes would call for making a decision, and economics had its input. Science generates knowledge; economics governs whether and how the new knowledge is used.

Today successful farmers hire (and pay well) young graduates of the schools of agriculture who are better versed than their elder employers in the new technology. And agribusiness firms hire young graduates of the schools of management who grew up with computers. Farmers get much help from agribusiness firms that supply them with their input items.

Today farmers employ consultants, bringing in expert judgment to help them make crucial decisions. Absentee owners employ farm management firms to supervise their tenants. Farmers buy the skills of these professionals because it is less expensive to do so than to develop these skills themselves. This is an application of the laws of comparative advantage and opportunity costs.

Agricultural research has become interdisciplinary. James Watson was a geneticist and Francis Crick was a crystallographer. Working together they discovered DNA, something neither could have done working alone. Norman Borlaug began as a pathologist. He moved into genetics, agronomy, herbicides, pesticides, economics, and engineering, hiring good people and learning from them the knowledge he lacked. Integrating these disciplines, he produced his high-yielding wheat. It is at the interface of scientific disciplines that new knowledge is born.

The old days of scientists working alone at their bench within a single discipline, as did Liebig and Pasteur, are fading away; team research is the new paradigm. Even basic lines of scientific inquiry are increasingly multidisciplinary. To be a specialist is great, but specialists should know enough about the other disciplines to understand what their colleagues are trying to tell them.

County agricultural agents have had to comprehend the results of interdisciplinary research; they cannot be specialists with a single area of knowledge. Farmers have had to learn how to apply knowledge; they must be skilled in a broad rather than a narrow sense.

Farmers are interested in the practical application of the diverse disciplines previously mentioned. They themselves integrate these disciplines at the farm level with the help of the economic principles mentioned earlier.

Chapter 7

THE AGRICULTURAL REVOLUTION IN A GLOBAL CONTEXT

The agricultural revolution greatly increased American capability to export farm products into the world market. Also, and more subtly, the American Agricultural Revolution paralleled a worldwide movement that substituted centralized decision making for the market system, agricultural and other. This trend was more evident in the Soviet Union and China but had its counterpart in the United States and other countries. In 1902, the federal, state, and local governments took 7 percent of the gross national product (GNP). By 1990, that percentage was 31. Agriculture reflected this trend; dissatisfaction with the outcomes of market principles outlined in Chapter 6 led to the transfer of much decision making from private institutions to government agencies.

FOREIGN TRADE IN FARM PRODUCTS

This chapter addresses the Agricultural Revolution as it is reflected in foreign trade in farm products.

The American Midwest is the garden spot of the world. From Columbus, Ohio, to Denver, Colorado, and from the Mexican to the

Canadian borders lies a huge area of agricultural endowment unmatched elsewhere else in the world, with fertile soil, good topography, generally adequate rainfall, and a climate conducive to plant growth. The area is laced with railroads, crisscrossed with highways, and serviced by great waterways. It has flourishing agribusiness firms and a good system of agricultural research and education. It is peopled by intelligent, hard-working farmers whose skills are unexcelled. There are, of course, other excellent agricultural areas in the United States: the central valley of California, parts of Florida, and sections of the Atlantic shore.

Thus endowed, the United States has an absolute advantage over most other temperate-zone countries of the world and is in good position to compete in the market in international trade.

During the 19th century, when the American Midwest was being opened up, the food-consuming population of the United States was small, and the channels of trade were reasonably free, a great flow of agricultural products moved from the United States to Europe, generating the foreign exchange with which the United States purchased the supplies needed for the American Industrial Revolution. From 1855 to 1860, agricultural exports were 82 percent of all American exports (USDA/FAS, 1953).

Like other entrepreneurs, farmers are anxious to sell their products abroad, so exports are very important. Also like other entrepreneurs, farmers object to imports that compete with their own products. Agricultural imports represent a loss of markets to farmers even though society as a whole is the beneficiary.

COMPOSITION OF FOREIGN TRADE IN FARM PRODUCTS

Chief agricultural exports today are oilseeds, feedgrains, wheat and flour, meat, and cotton. Chief agricultural imports are coffee, animals and animal products, sugar, vegetables, and rubber.

Imports of farm products may be divided into two groups: those that compete with our crops and livestock and those that do not. Competing products are animals and animal products, sugar, oilseeds of various kinds, vegetables, wines, tobacco, fruit, nuts, and certain grains and wool. Those that do not compete—chiefly those from outside the temperate zone—are coffee, rubber, bananas, cocoa, drugs, spices, and tea. Competing products have a value about twice that of those that do not compete. Table 7-1 shows total agricultural exports, imports, and agricultural trade balances by decades for the 20th century. The table shows the

TABLE 7-1. U.S. agricultural exports, imports, and trade balances for selected periods during the 20th century.

Period	Exports	Imports	Balance
	(billion dollars)		
1915–1919	2.6	0.8	1.8
1920–1929	1.9	1.0	0.9
1930–1939	0.8	1.1	-0.3
1940–1945	1.5	1.6	-0.1
1946–1949	3.6	2.8	0.8
1950–1959	3.6	4.2	-0.6
1960–1971	6.1	4.5	1.6
1972–1981	26.6	12.4	14.2
1982–1988	32.7	19.7	13.0
1989–1998	47.3	27.9	19.4

Sources: 1915–1940: USDA, Foreign Agricultural Service, 1953, *United States Farm Products in Foreign Trade*, Statistical Bulletin No. 112, Washington, D.C.: USDA/FAS; 1940–1985: Council of Economic Advisers, *Economic Report of the President*, 1987, Washington, D.C.: U.S. Government Printing Office; 1985–1998: USDA, Economic Research Service, various issues, *Agricultural Outlook*, Washington, D.C.: USDA/ERS.

phenomenal increase in agricultural trade during the last quarter of the 20th century and the favorable agricultural trade balance of that period—a reflection of the magnificent agricultural resources of the United States and growing impact of the Agricultural Revolution.

In the first decade of the 20th century, 15 to 25 percent of farm cash receipts from marketing were earned in export markets (USDA/FAS, 1953). During these years tariffs were a component of economic policy. On the whole, farmers favored tariffs, though their interests were not served by them. Tariffs added to the prices farmers had to pay for machinery and such goods while giving them little protection for the products they sold.

In 1909 Congress passed the Payne-Aldrich Act, which placed duties at the highest level up to that time. A tariff is an interference with international trade, but it is not a specific obstacle governing individual transactions, as is a quota.

Prior to the 1920s, farmers had pursued public policies of general application, such as monetary policy and tariffs. During the 1920s, and especially during the 1930s, they sought—and obtained—government actions that interfered directly with individual decision making.

The years before World War I, 1910 to 1914, became known as the Golden Age of agriculture. Farmers emerged from the depressed years of the late 19th century. The rising general price level increased prices of articles that farmers sold, while the prices they paid lagged behind. The World War I period was a time of great agricultural prosperity. "Food will win the war" became the accepted slogan. Vast areas of the Great Plains were plowed up to produce wheat for export. The level of farm prices more than doubled. The price of farm land rose by 70 percent between 1913 and 1920 (Cochrane, 1979). Euphoria prevailed throughout rural America.

Herbert Hoover was named food administrator by President Woodrow Wilson. American food supplies were generally abundant throughout the war. Hoover avoided general rationing and stressed food conservation. If he was a "food czar," as some claimed, he was a benevolent despot.

The war ended in 1918. For a brief time the momentum of high farm prices continued as the United States fed war-ravaged Europe. But then came disaster. As European farm production recovered, products of the acreage brought under production to feed Europe could no longer be exported at high prices. Exports shriveled. The price of wheat fell from over $2.00 to under $1.00 a bushel. To complicate matters, the newly created Federal Reserve System, alarmed about inflation, curtailed credit, raising the discount rate by three percentage points in a seven-month span during 1919 and 1920. With the deprivation of money—the lubricant that facilitates trade—prices collapsed. From 1920 to 1922 the general price level fell 38 percent. A long period of farm distress followed. For farmers, the Great Depression began in 1921; the 1929–1933 episode was disaster on top of distress.

The slogan in farm areas was "Make the tariff effective for agriculture." Falling farm prices led to the idea of insulating farmers from falling world prices by one means or another. The export market was seen as an opportunity to unload surpluses abroad. Two-price plans were proposed: a high price for what was sold in the domestic market and a low price for what was sold abroad. Farm politicians avoided the word "dumping," but that is what was being proposed. This proposal was supported by farmers and was enacted by Congress but was twice vetoed by President Calvin Coolidge.

In 1930 Congress passed the Smoot-Hawley Tariff Act, a major intrusion into what otherwise would have been reasonably competitive trade. Tariffs were raised to an all-time high (Benedict, 1953).

Farmers generally supported this act, not realizing that to impose a tariff on an exported crop would have no effect.

International trade, already in retreat because of the onset of the Great Depression, stagnated. A thousand economists had implored President Hoover not to sign the Smoot-Hawley tariff bill, but he signed it nonetheless. The Smoot-Hawley Tariff Act must bear some responsibility for the shrinking of foreign trade and the Great Depression that followed. The Great Depression was, for agriculture, the defining event of the 20th century.

As the Great Depression deepened, officials of the Federal Reserve System, through fear of inflation, declined to provide the financial means by which normal trade could occur. The supply of money fell by one third. Prices received by farmers fell. From 1929 to 1932, prices received by farmers declined 56 percent. Gross farm income in the United States fell 54 percent. Net farm income fell from $6.3 to $1.9 billion. Strapped for cash, farmers stopped buying. The wry farm comment was as follows:

Use it up,
Wear it out,
Make it do,
Or do without.

The Agricultural Revolution slowed to a creep.

In 1933 Congress set up the Agricultural Adjustment Administration (AAA), curtailing farm production and raising prices received by farmers. Reduced production and increased prices curtailed foreign trade in farm products. The great natural advantages enjoyed by American agriculture, described in the opening paragraph of this chapter, were foregone. Exports of farm products suffered a double blow: reduced farm production and American prices held above the world level.

Nowhere was the deleterious effect of restricted production more visible than in the case of cotton. With our prices increased and our exports limited to our status as a market of last resort, we held an umbrella under which cotton production increased in India, Brazil, and the Soviet Union (Table 7-2).

CLASSIC FAILURES OF FARM POLICY

F.A. Pearson, a professor at Cornell University during the 1940s, listed three classic failures of farm policy, all of which were the result of increasing prices above the competitive level:

TABLE 7-2. Production of cotton, domestic and foreign, before and after passage of the Agricultural Adjustment Administration (AAA).

Production	Before AAA 1928–30	After AAA 1976–77
	(millions of bales)	
United States production of cotton	14.4	12.6
Foreign production of cotton	12.2	49.2

Source: USDA *Agricultural Statistics,* various issues.

1. The Malayan program to boost the price of rubber, which put neighboring Indonesia in the rubber business.
2. The Brazilian program to increase the price of coffee, which put Africa into coffee production.
3. The American effort to raise the price of cotton, which held the price umbrella for expanded production in the rest of the world.

CLASSIC FARM POLICY SUCCESSES

Offsetting these classic farm policy failures were various American farm policy successes, including the following:

RESTRUCTURING FARM CREDIT

In the old days, a farmer would go to his bank and ask for credit, say to buy a farm. The banker would grant him credit at a high rate, perhaps for five years, less time than would permit repayment from the proceeds of the loan. At the end of five years, the farmer would appear, hat in hand, asking for an extension of the loan, which the banker might or might not grant. If the banker was hard-hearted, he might foreclose.

With the help of two Cornell professors, F.F. Hill and W.I. Myers, these numerous small farm loans were consolidated and offered in huge chunks in the New York money market, producing large sums of money that permitted loans to farmers for 30 instead of five years, at competitive interest rates, through the Farm Credit Administration (FCA). Murray Benedict (1953) tells how, during the hard times of the 1930s, thousands of farms were thus saved from foreclosure. The senior author's family farm was one of these.

International Research Network

This system (further described in Chapter 8) was the institutional invention of a number of people, among them David Bell, F.F. Hill, Lowell Hardin, George Harrar, Robert M. McNamara, and Sterling Wortman. The Rockefeller Foundation had a strong hand in this initiative, which permitted the concentration of various agricultural disciplines (agronomy, biology, chemistry, economics, genetics) on special problems of particular crops and livestock. It produced such marvels as high-yielding wheat and miracle rice, which saved much of the Third World from disaster during the failed crops of the 1960s and 1970s.

Public Law 480 (Food for Peace)

This law was the brainchild of Gwynn Garnett of the American Farm Bureau Federation. Garnett conceived the idea of moving American surplus wheat, (which was costly to store and for which there was no commercial outlet) to hungry foreign nations. The wheat was paid for not in hard currencies (of which they had little) but in their own currencies (of which they had an abundance). This was a thinly disguised "sale," which offended those devoted to international trade in the conventional manner. But it helped solve the twin problems of costly unsalable surplus in the United States and food deficits in the Third World. This episode is reported briefly by the senior author (Paarlberg, 1997). Major technological changes create the need for institutional innovations and new human capital.

These three institutional inventions, all of the 20th century, were comparable to (but of an utterly different nature) the innovations in biology, chemistry, and engineering named earlier in this book. But they are equally deserving of public acknowledgment.

War Begins, Great Depression Ends

World War II began in 1939 and broke the back of the Great Depression. Claude Wickard, Secretary of Agriculture, revived and extended the slogan of World War I. He said, "Food will win the war and write the peace." Farm production was encouraged and agricultural exports rose. Prices rose above the price floors established by farm legislation and even above price ceilings voted by Congress to suppress inflation. Production controls were relaxed. But the production-restricting, price-supporting legislation was kept in place; there was widespread apprehension of an economic collapse after World War II, as had happened after World War I, and farmers felt that the old program of government intervention would be needed.

But this anticipated price collapse did not occur. Incomes saved during the war were spent during the postwar period, strengthening demand. Wise macroeconomic policies on the part of the Federal Reserve Board averted disaster. Monetary actions (or inactions) that led to the disasters of 1920 and 1929 were avoided. Keynesian principles and astute centralized financial decision making were the economic triumphs of the postwar world.

Managed Trade

During the middle of the 20th century, exports of United States farm products fell to only 10 percent of income from farm sales, which was not surprising since we had reduced the supply of such products and priced them out of the foreign market. After World War II we moved into a system of governmentally managed international trade, including trade in farm products. International agreements of varied governmental discipline were invoked for wheat, cotton, and other products.

The General Agreement on Tariffs and Trade (GATT), enacted in 1947 and updated in 1994, liberalized international trade to some degree in many kinds of products, including (belatedly) agriculture. The North American Free Trade Agreement (NAFTA), a liberalizing action, was passed in 1993. Other government interventions occurred. Export subsidy programs of one kind or another were adopted by the United States. Other interferences with foreign trade include bilateral trade agreements, market promotion and development, and lending to foreign countries for the purchase of farm products. Export embargoes were imposed in 1973 in response to greatly increased farm prices (allegedly to avoid drawdown of American stocks of oilseeds) and again in 1974 and 1975, when lower United States crops were forecast. In 1980 sales to the Soviet Union were canceled in response to Soviet invasion of Afghanistan. These embargoes were generally opposed by the farm community and were questionable in their effect.

During the 1970s there was apprehension about whether the world could feed its people. Poor crops worldwide gave fuel to this concern. It was thought by some that American exports of food would be needed to save the world from starvation. Prices of farm products rose 120 percent. Farmers, believing that food shortage was in prospect, borrowed money to buy land and equipment. Prices surged upward to double-digit levels of inflation, not just in agriculture but generally. In 1979 the Federal Reserve System, fearful of inflation, curbed credit and by 1981 boosted the interest rate to 19 percent. It slammed on the brakes, ignoring the disasters of 1920 and 1929. It threw farmers against the windshield.

The alleged shortage did not materialize. Production increased. Higher prices restrained consumption. Farm prices fell sharply. Exports shriveled. In 1981 exports of farm products totaled $43 billion; in 1986, they were $26 billion. Farmers were hit with a triple whammy: soaring interest rates, falling prices, and shriveled exports.

The poor export performance caused the revival of the export assistance that had been reduced during the 1970s. Export credit funding was increased. PL 480 was expanded, old export subsidy programs for bulk commodities were revived, and new subsidy programs for value-added and processed goods were started. By the late 1980s the U.S. government was again heavily involved in managing exports.

CHANGING COMMODITY COMPOSITION OF U.S. TRADE

In a dynamic environment like the global agricultural economy, a key factor for a country to maintain its trade position is its flexibility to shift production as global demand changes. Agriculture in the United States has shown considerable production and trade flexibility during the 20th century.

During the 1910–14 period cotton was the largest U.S. agricultural export category on a value basis just as it had been since Colonial times. Indeed, cotton exports accounted for roughly half of total U.S. agricultural exports. In second place, though a distant second, were wheat and flour exports, at 13 percent of U.S. agricultural exports. Third place was held by animal fats, with meat and products a close fourth. All other grains and products at about 5 percent were in fifth place. Tobacco and oilseeds (largely cottonseed) competed for the next spot.

By the end of the 20th century these rankings had shifted. Oilseeds and their products were the dominant U.S. agricultural exports at about 20 percent of total exports. Whereas this category had consisted of cottonseed at the start of the century, by the century's end the category was largely soybeans and soybean products. Cotton, the early king, accounted for only about 5 percent of U.S. agricultural exports in 1998. Wheat, too, had declined in importance, though not as badly as cotton. Exports of other grains, such as corn, barley, rye, and rice, which had been of little importance in 1910, were the second most important in 1998.

The shift in the commodity composition of U.S. agricultural exports started early in the 20th century. During World War I the food needs of the Allies shifted U.S. exports away from cotton in favor of wheat, other grains, and meat. The global economic expansion following World War II accelerated changes in the composition of U.S. agricultural exports. Economic recovery and income growth caused dietary changes that propelled grain and oilseed exports to prominence. Increasing incomes in higher-income nations boosted the demand for meat and thus for feed grains as well as protein meals made from oilseeds. Increasing incomes and urbanization in lower-income nations also shifted food consumption away from traditional foods (such as rice and potatoes) toward bread made from imported wheat.

This evolution continued. Until around 1990 bulk commodity exports (grains and oilseeds) dominated U.S. agricultural exports. After 1990, rising incomes around the world boosted demand for more processed and value-added goods. The most rapidly growing U.S. agricultural exports during the 1990s were meats, horticultural products, fruits, and consumer-ready goods.

Imports of agricultural goods have changed as well over the 20th century. During the early decades, U.S. imports were dominated by three commodities: rubber, coffee, and sugar. Rubber could not be grown in the United States, so the demands for tires, belts, and such had to be met by imports. Coffee also could not be grown in the United States, so it was imported. Some sugar was produced domestically, but that production was insufficient for the quantities demanded by Americans. Other goods—fruits, vegetables, grains, meats, and vegetable oils—were imported, but the values were small compared to the big three.

The end-of-century import picture differed greatly. The largest import category for the United States was fruits, followed by vegetables. Coffee remained an important import, but its third-place ranking was challenged by grains and by meats. The changes in import composition reflected the increase in U.S. per-capita incomes and changes in diets. Modern Americans consumed more fruit and vegetables. Modern Americans also expanded consumption of foods from other countries (such as Mexico, Italy, Greece, China) that required imported ingredients.

NEW DESTINATIONS, NEW SOURCES

In addition to a change in the commodity composition of U.S. agricultural exports, the destinations of those exports changed. That process continued throughout the 20th century and will continue into the 21st century.

In the first decades of the 20th century, U.S. agricultural exports were focused on Europe. For the 1910–1914 period, Europe bought 85 percent of all U.S. agricultural exports, and Great Britain, the largest single buyer, 37 percent. Australia, New Zealand, and Canada bought 6 percent. Asia, Latin America, and Africa combined purchased a mere 9 percent.

The U.S. agricultural export data for 1998 show a picture typical of the later years of the 20th century. Europe bought around 19 percent; Asia, 45 percent; and Latin America, 18 percent. The largest single buyer was Japan (18 percent), followed by Canada (12 percent), and Mexico (9 percent). Britain, which started the century as the most important destination, ended the century as the 10th largest buyer.

While the shift away from Europe as a destination for U.S. agricultural exports was a continuous process over the century, three distinct turning points can be identified.

The first occurred during the Depression and World War II, when U.S. agricultural exports began to diversify their destinations. Exports began to be shipped to South Asia (India and Pakistan), the Near East, and Latin America. European countries remained the top markets but their shares fell.

The second turning point occurred in the middle 1970s. Those years were characterized by extreme global production shortfalls in the Soviet Union, India, and elsewhere. The United States helped to meet them through expanded exports. This was the period when the importance of exports for U.S. agriculture reemerged. Trade destinations shifted toward centrally planned nations and the developing countries. Nations in Asia, North Africa, and the Middle East became critical markets for U.S. agriculture. Western Europe reversed its trade pattern and became an export competitor of the United States in many commodities.

The third turning point occurred in the late 1980s and early 1990s. This period was associated with rapid income growth and economic reforms in Asia and Latin America. Trade in feedstuffs and high-value goods increased in importance. A process of regional trade integration began with Mexico, Canada, and the nations of the Pacific.

As the century draws to a close, the dynamism of the Agricultural Revolution is being confirmed. The expectation is that U.S. agricultural exports will continue the process of shifting toward developing nations in Asia, like China, and toward closer links with Latin America. The forces driving this process include the increased capability of American agriculture, coupled with economic reform abroad, which will raise living standards and increase the demand for meats and feedstuffs that the United States can supply at low cost. Another force is the tendency for regional trade agreements with these nations to foster market integration.

A threat to exports of American farm products to Europe is consumer concern about "genetically modified organisms" (discussed briefly in Chapter 4).

Late in the 1990s economic decline occurred in southeast Asia, resulting in diminished agricultural exports to that area. The prevailing assessment is that this falloff is temporary and that, given time, former trends will be restored.

With the changes in the import composition, there also was a change in where imports were obtained. Rubber, one of the big three imports in the early years, came from Malaysia, the Philippines, and Indonesia. Coffee imports came from Brazil and Central America. Before the Cuban revolution, sugar was supplied mostly by Cuba. Latin America is still a dominant supplier of coffee, fruits, and vegetables. The rise in fruit and vegetable imports has increased the importance of Mexico as a supplier. Cuba no longer supplies sugar to the United States; that role is performed primarily by the Caribbean nations and the Philippines. The jump in meat imports has raised the importance of Australia, Canada, New Zealand, and Mexico to U.S. agricultural imports. Similarly, the rise in the share of grains in the U.S. import picture has boosted the role of Canada. The potential of the Agricultural Revolution is portrayed in these changing patterns.

The 20th century record of international trade in farm products reflects American concern about alternating shortage and excess of farm products, efforts to reward friends and punish adversaries, altering patterns of political power, and the explosive force of the Agricultural Revolution.

During the last quarter of the century, agricultural exports, measured as a percentage of receipt of sales of farm products recovered substantially (Table 7-3).

SUMMARY

The export trade in farm products is not the first claimant on food supplies when domestic markets get tight, and it is the preferred disposal outlet when domestic supplies are abundant. Given this circumstance, volatility in the export market for farm products, both for price and volume, is unavoidable.

The Great Depression of the 1930s and the associated collapse in foreign trade slowed the Agricultural Revolution to a near halt. The emergence from the Depression and the thriving markets associated with World War II speeded it up. For the greater part, farmers made decisions based on their perceived best interests, and the Agricultural Revolution advanced in irregular fashion.

TABLE 7-3. Agricultural exports as a percentage of product sales.

Years	Percent
1900–1925	15–25
1926–1975	10
1976–2000	25–30

Sources: 1900–1950: USDA, Foreign Agricultural Service, 1953, *United States Farm Products in Foreign Trade,* Statistical Bulletin No. 112, Washington, D.C.: USDA/FAS; 1951–1998: USDA, various issues, *Agricultural Statistics,* Washington, D.C.: U.S. Government Printing Office; 1999–2000: USDA, Economic Research Service, December 1999, *Agricultural Outlook,* Washington, D.C.: USDA/ERS.

A field gang weighs in the harvest, which involves stoop labor and piecework. (Photo: USDA)

Tying a shock of corn (1917). Corn was cut and set in a shock for the ears to dry. The dried corn was first bound with a rope to pull the stalks snug, and then the shock was tied with twine and the first rope removed. Later, the stalks with the ears were either brought to the barn for husking and shredding, or the ears were husked in the field and the stalks hauled to the barn—a labor-intensive harvest. (Photo: J. C. Allen and Son)

Farm power, all forms: touring car, mules, tractor, and truck. (Photo: J. C. Allen and Son)

Husking corn in the field (1918). Workers tossed the husked corn into the wagon without looking; the corn hit the "bang board" and dropped onto the accumulating pile. Good workers claimed they could "keep an ear in the air." The ultimate mark of skill was husking and unloading 100 bushels a day. (Photo: J. C. Allen and Son)

A two-horse team pulls a tractor out of a soft spot in a sod field. (Photo: J. C. Allen and Son)

Shelling corn with steam power out of an old-style corn crib. (Photo: J. C. Allen and Son)

Plowing with an old-fashioned tractor. The disk in the furrow automatically did most of the steering. (Photo: J. C. Allen and Son)

Planting cabbage on the home farm in Lake County, Indiana (1930s). Senior author, seated, far left.

A one-row, one-horse, two-man planter. (Photo J. C. Allen and Son)

This family of 11 took pride in picking more than a bale of cotton (500 lb) in a day. With a modern four-row, mechanical cotton picker, one person can now harvest 80 bales a day. (Photo: USDA)

Tractors converted from Model T Fords: one plowing and the other pulling a spring-tooth harrow. (Photo: USDA)

Aerial spraying of a young orchard. (Photo: USDA)

A six-horse team pulling a harrow. (Photo: USDA)

Hybridizing wheat (1968): a delicate operation using tweezers. Bags on wheat heads keep out foreign pollen. (Photo: Purdue University)

Disking and drilling grain in a single pass over the land with a Case tractor (1934). (Photo: J. C. Allen and Son)

A big horse hitch pulls a three-bottom plow (1927). (Photo: J. C. Allen and Son)

Threshing soybeans with a thresher powered by a Huber tractor (1928). (Photo: J. C. Allen and Son)

Soybeans in corn residue in Jackson County, Illinois: minimum tillage. (Photo: USDA, Gene Alexander)

No-till planting of wheat in Whitman County, Washington. (Photo: USDA, Tim McCabe)

No-till planting of corn near Plymouth, Iowa. (Photo: USDA, Gene Alexander)

A worker, wearing gloves, mask, and protective clothing for safety, empties Roundup herbicide into a storage tank. (Photo: USDA)

Spraying corn for insect control. The corn is being hybridized; that is, two male rows in tassel are alternated with six detasseled female rows. (Photo: USDA)

Pivot irrigation on cotton in Mississippi. (Photo: USDA, Tim McCabe)

Contour farming. (Photo: USDA)

Harvesting barley in Colorado. (Photo: USDA, Gene Alexander)

Harvesting peas near Rio, Wisconsin. (Photo: USDA, David F. Warren)

Wheat harvest in El Centro, California. (Photo: USDA, Tim McCabe)

A modern cotton picker near Mound Bay, Mississippi. (Photo: USDA, Bill Tarpenning)

Hand-picking apples. (Photo: USDA)

Mechanical harvesting of tomatoes. (Photo: USDA)

Chapter 8

EXPORTING THE REVOLUTION

Revolutions—whether political, industrial, or agricultural—have a way of spreading. So it was with the Agricultural Revolution of the 20th century. This chapter records efforts to augment the spread of that revolution in terms of both product and technology into areas outside the United States, chiefly in what we call the Third World. We call it "exporting the revolution."

We exported the results of the American Agricultural Revolution by sending abroad during the last quarter of the 20th century approximately 25 percent of our agricultural production.

We exported the Agricultural Revolution by means of the following:

- Educating in land grant agricultural colleges during the school year 1997–98 8,570 agricultural students from foreign lands, and preparing during the school year 1996–97 1,165 American agricultural students for service abroad.
- Contributing $38 million in 1998 to the Consultative Group on International Agricultural Research (CGIAR), the group that finances the International Research Network.

- Contributing $15 billion in 1991 for international agricultural development through the federal government, mostly the U.S. Agency for International Development (USAID) (about which more is reported later in this chapter). Most polls indicate that Americans think foreign assistance is the largest single item in the federal budget. Actually this item amounts to only one half of one percent of the budget.
- Sending abroad American farmers and scientists from our land grant colleges to encourage better farm practices in the Third World.
- Contributing privately to voluntary organizations (PVOs) and nongovernment organizations (NGOs) such as the United Nations Children's Fund (UNICEF), Cooperative for American Relief Everywhere (CARE), Catholic Relief Services, Church World Service, International Confederation of Catholic Organizations for Charitable and Social Action [*Vatican*] (CARITAS), and United Methodist Committee on Relief (UMCOR), which combine agricultural betterment abroad with the relief of human distress.
- American farm equipment firms such as Caterpillar, Case, John Deere, and New Holland. These firms built enclaves of agricultural improvement in the less-developed countries, which then expanded and grew. In similar fashion, British firms invested capital in our agriculture during the early days of this country, laying the foundation of the American Agricultural Revolution.
- Research done by American land grant colleges and great American firms such as Monsanto and DuPont.

International Research Network

Perhaps the most dramatic export of the of the Agricultural Revolution is the International Research Network. Sixteen centers of this network are located throughout the developed and the less-developed world and are coordinated by the Consultative Group on International Agricultural Research (CGIAR) in Washington, D.C., founded in 1971. They are listed below. Those created before 1971 were founded and funded by the Rockefeller and Ford Foundations.

CIAT, (Centro Internacional de Agricultura Tropical), the International Center for Tropical Agriculture, Cali, Colombia, South America, founded in 1967 and focusing on beans, cassava, tropical forages, and resource management in human ecosystems.

CIFOR (Center for International Forestry Research), Bogor Barat, Indonesia, founded in 1992, focusing on forest systems and forestry.

CIMMYT (Centro Internacional de Mejoramiento de Maiz y Trigo), the International Center for Improvement of Corn and Wheat, Lisboa, near Mexico City, Mexico, founded in 1966, where Norman Borlaug developed his high-yielding wheat, which helped avert mass starvation in the Third World.

CIP (Centro Internacional de la Papa), the International Potato Center, Lima, Peru, founded in 1971, focusing on potato and sweet potato.

ICARDA (International Center for Agricultural Research in the Dry Areas), near Aleppo, Syria, founded in 1977, focusing on the improvement of agricultural systems involving barley, lentils, fava beans, wheat, chickpea, and small ruminants (sheep and goats).

ICLARM (International Center for Living Aquatic Resources Management), Makati City, Philippines, founded in 1977, focusing on the improvement of the production and management of aquatic resources (fish and fisheries).

ICRAF (International Centre for Research in Agroforestry), Nairobi, Kenya, founded in 1977, focusing on the improvement of agroforestry systems, with trees and cultivated crops in the same field.

ICRISAT (International Crops Research Institute for the Semi-Arid Tropics), Andhra Pradesh, India, founded in 1972, focusing on sorghum, finger millet, pearl millet, chickpea, and groundnut.

IFPRI (International Food Policy Research Institute), Washington, D.C., founded in 1975, focusing on identifying and analyzing alternative national and international strategies and policies for meeting food needs.

IITA (International Institute of Tropical Agriculture), near Ibadan, Nigeria, founded in 1967, focusing on cassava, corn, cowpea, soybean, yam, banana, and plantain.

ILRI (International Livestock Research Institute), Nairobi, Kenya, founded in 1995, by the merger of the International Laboratory for Research on Animal Diseases (ILRAD, founded in 1974) and the International Livestock Centre for Africa (ILCA, Addis Ababa, Ethiopia, also founded in 1974).

IPGRI/INIBAP (International Plant Genetic Resources Institute/International Network for the Improvement of Banana and Plantain, affiliated with CGIAR), Montpellier, France, founded in 1974, focusing on strengthening the conservation and use of plant genetic resources worldwide.

IRRI (International Rice Research Institute) Makati City, Philippines, founded in 1960, where the so-called miracle rice was developed, which, with Borlaug's high-yielding wheat, created the Green Revolution and helped avert disaster during the failed crops, worldwide, of the 1970s.

ISNAR (International Service for National Agricultural Research), The Hague, Netherlands, founded in 1979, focusing on helping national agricultural research systems and organizations improve their performance.

IWMI (International Water Management Institute), Colombo, Sri Lanka, founded in 1984, focusing on the improvement in the management of water resources and irrigation systems.

WARDA/ADRAO (West Africa Rice Development Association/Association pour le Developpement de la Riziculture in Afrique de l'Ouest), Bouaké, Ivory Coast, founded in 1970, focusing on the improvement of rice crops.

These centers are supported by an international consortium of 50 donor members (20 industrial nations, 16 developing countries, 11 international and regional organizations, and 3 foundations that encourage agricultural betterment in Third World countries). Since the beginning in 1971, there has been a public investment of $4.7 billion in this network. It employs some 800 scientists.

The International Research Network focused naturally on research. Originally, the network addressed particular crops, such as wheat, rice, and corn. But that is changing; the paradigm has shifted, as shown in Table 8-1 (Hardin, 1998).

The International Research Network, of which the CGIAR-associated centers are a part, is but one of the research enterprises that focuses on agricultural development abroad. USAID and other pub-

TABLE 8-1. Paradigm shifts in crop research

From	Toward
Food supply focus	Food access
Classical agronomy	Ecological sciences
Analytical research	Systems dynamics
Top down	Participation
Factor-oriented management	Integrated natural resource management
Public goods	Private ownership

Source: Adapted by L.S. Hardin, 1998.

lic agricultural research agencies in the United States and overseas support international agricultural research and improvement. The developing countries themselves, out of their limited resources, have spent more than half of the outlay.

Other great American agencies promote agricultural development abroad. Winrock International, headquartered in Arkansas and Washington, D.C., has a program to bring lasting and positive change to people around the world. In 1997 its budget was over $30 million.

Global 2000, Inc., a nonprofit organization headed by former U.S. President Jimmy Carter, headquartered in Atlanta, Georgia, has been working with more than 600,000 small-scale farm families in at least a dozen countries in Africa. Norman Borlaug, the American agricultural scientist and Nobel laureate, has led this effort to improve crop yields, working with heads of state and their ministries of agriculture.

Agricultural research is not the only initiative undertaken in an effort to achieve human betterment in the Third World. Providing research data to operating farmers, development of infrastructure, and institutional reform all give balance to the initiative.

EXPORTS FROM OTHER COUNTRIES

Lest we think exclusively in terms of what we do to help other countries, consider what other countries have done to help us give new life to our Agricultural Revolution.

The great pioneers who laid the basis for the Agricultural Revolution—chemists, biologists, economists—were all European.

We imported the idea of a publicly supported Agricultural Experiment Station from Germany. We currently import agricultural machinery from Europe, Japan, and Russia. We import agricultural products such as wool, for which other countries have a comparative advantage, thus serving American consumers better than if they were to rely on higher-priced domestic products for which American farmers are at a comparative disadvantage.

No important crop or livestock breed was domesticated in what is now the United States. Almost all the crops and livestock we produce originally came from elsewhere. Wheat, originally domesticated in western Asia, came to the United States by a variety of routes, most recently from Russia. Corn came from Central America, soybeans from China, rice from Asia, potatoes from the Andes, and sorghum from North Africa. Cotton appears to have been domesticated in both the old and new worlds.

With livestock it is no different. Our beef cattle breeds came mostly from the British Isles. The Santa Gertrudis breed of cattle has blood lines from India. Our dairy cattle came from the Netherlands and from the British Isles. Our sheep came from a number of countries, including Spain. Our hogs were mostly British breeds. Our horses came from Spain, from the British Isles, and from Arabia.

The idea that American agriculture ever was or can be confined to the United States is simply incorrect.

CAPITAL INVESTMENT

The world has grown weary in well-doing. Total bilateral and multilateral assistance to agriculture in developing countries was $10.3 billion in 1995, down 20 percent from 1991 aid levels.

The people involved early in the USAID program struggled for a strategy that would work. They tried the idea that had worked in the Marshall Plan—"capital investment." Europe had an educated, highly motivated populace, who had in mind the functioning economic structure of prewar times and needed only the restoration of war-destroyed capital. In the Third World, there was no such remembrance and no such vision. The leap from the oxcart to the tractor was greater than was readily possible for people who had no such institutional memory. After all, it had taken better-educated American farmers nearly 100 years to make a change of like magnitude.

We tried "incremental change," which was better than capital investment. We tried crash programs, some of which crashed. We tried "big push," which generally proved to be too big. We developed "two-sector models," which treated farm and nonfarm problems differently. "Balanced growth" was the paradigm for a while. "Small is beautiful" had its day. "Get the prices right" was the slogan for a time. We set out to "help the poorest of the poor." "Food first" emphasized the agricultural sector relative to industry. "Intermediate technology" was fashionable for a while. "Import substitution" was briefly in vogue, followed by "export-led growth."

We did not adopt the "lifeboat strategy," which was offered by the Paddock brothers in 1967. The idea was that the world had outrun its food supply. Figuratively, the ship had sunk and the lifeboat was full of people. If more people struggled to get into the lifeboat, it would sink and all would be lost.

A similar strategy, called "triage," was offered by Garrett Hardin in 1974. The word *triage* comes from a World War I policy. The battlefield would be strewn with too many wounded soldiers for the medical corps to care for. The injured would be divided into three groups: those who could care for themselves at least for a time, those who could respond favorably to the medical care that was available, and those who were beyond help and would be left to die.

Such was the case in the world, said the Paddock brothers and Garrett Hardin during the world food scare of the 1960s and 1970s. Some countries, like India and Bangladesh, they said, should be written off as beyond hope. Others, like Brazil, could make it on their own. Our limited resources should be provided only to those countries in which help could make the critical difference.

It was during this terrible time that there appeared, in 1972 and again in 1974, the gloomy *Report to the Club of Rome*, by Donella Meadows et al., that envisioned a date when various disasters would coincide: the population explosion, the degradation of resources, deteriorating crop yields, and intolerable pollution. Anyone who was ethically sensitive experienced anguish during that time.

Thanks to the Agricultural Revolution, this worldwide disaster did not happen. The gloomy prognosticators made a grievous error—they underestimated the dynamism of the Agricultural Revolution.

With Point Four (agricultural development, technical assistance), we were moved by a mix of purposes:

- To help unfortunate people.
- To win friends abroad and the belief that our programs would help us do so.
- To build the developing nations into good trading partners.

- To provide technical assistance that would help meet aspirations of the people and so promote peace.
- To counter the threat of communism in the Third World by offering a better alternative: democracy with its economic progress. (When the Soviet Union collapsed in 1992, this later motive atrophied, and our international agricultural initiative lost a large part of its motivation.)

FAIL-SAFE AS AN EXPLANATION FOR AGRICULTURAL STAGNATION

In 1964, T.W. Schultz, a professor at the University of Chicago, produced his small book *Transforming Traditional Agriculture*, which won him a Nobel prize. Schultz, a farm boy from South Dakota before he became a professional economist, wrote that farmers—even primitive farmers—are wise in the use of their resources, having learned over the generations how to make best use of the land and capital available to them within the limits imposed by their education and experience. They are slow to venture, said Schultz, because a new initiative might fail. For them the traditional practice had proved its worth; it had helped them survive. What better measure of success? Thriving American farmers can take a risk. "Fail-safe" to them is a comprehensible thing. To farmers on the edge of want, risk is to be averted; failure may mean starvation.

Schultz's idea prescribes research and education as central to agricultural development in the Third World, as a means of positively disrupting the traditional equilibrium. Imparting agricultural knowledge directly to farmers through the extension service will work only when there is something worthwhile to extend. This requires innovative research, with dissemination and adoption of its findings. This simple prescription for agricultural development was understandable to American farmers, who could trace the progress of American agriculture by this same pattern.

What have we learned in our 50-year effort to export the American Agricultural Revolution to Third World countries? We have learned that research and education work, though slowly. Incentives work. F.F. Hill, former vice president of the Ford Foundation, said, "Farmers in the developing countries may not be able to read, but they know what pays."

The International Research Network works. It is voluntary, coordinated, interdisciplinary, and mission oriented. Some people find a supposed quarrel between the International Research Network, which is mission oriented, and the land grant college system, which is disciplinary. Which is better? Actually, they need each other—the land grant college system to produce excellent agricultural scientists and the International Research Center to put them to work. Various studies report that public investment in agricultural research has earned returns as high as 80 percent.

We have learned that the basic agricultural sciences—genetics, pathology, and plant nutrition—are applicable, but they must be suited to local needs. American practices must be *adapted*, not *adopted*.

We have learned that some things do not work. We have learned that it is best not to send overdeveloped scientists to underdeveloped countries. American farm technology cannot be exported as such. American methods of production suited to a 2,000-acre Iowa farm cannot be exported to a three-acre wheat farm in India, where the operator is uneducated and without the power to conceptualize what might be.

We Americans are inclined to think that farmers in the Third World should welcome with open arms the help that we so generously offer. But that is not always so. The recipients say our efforts are almost exclusively production oriented, while their problems are not only production-related but also social and political. They say our efforts are an extension of the American system—an effort to export capitalism as it exists in the United States and to prop up institutions that the reformers wish to overthrow. They contend that we promote energy-intensive and labor-saving agriculture; they lack energy and have superabundant labor.

So our effort to export our Agricultural Revolution to the less-developed world ran into grave problems. Opponents in the United States say, "It is no business of ours." "It doesn't work." "We put our rivals in business." "It isn't needed." "We upset an adopted culture." All of these arguments have some validity. They all seemingly justify inaction. One is reminded of the old adage, "All that is required for disaster to occur is that good men should do nothing."

What can we legitimately say about exporting the American Agricultural Revolution? The less-developed countries need what we can intelligently supply. Whether we will do so on an adequate scale is in doubt. The internationalists favor it and the nationalists oppose it.

Fifty years is not enough time to establish or to refute the principle of international interdependence. National interdependence is still on trial in the United States. There is doubt about extending this principle beyond our own borders. But time is probably on the side of a broadened rather than a narrowed feeling of responsibility.

International agricultural development is not a zero-sum game as some claim it to be. It is a positive-sum game with potential gains for all. The United States gains by bringing back to this country some of the good germplasm from other countries and by learning the value of multiple cropping and of aquaculture. In addition, we solve some pest and pathogen problems before they reach our shores. The Agricultural Revolution contains a growth factor. Just as it grew within our own land, so it is likely to grow in the less-developed world.

Supporters of our efforts to export the Agricultural Revolution point out that agriculture has advanced in the countries we have helped. Some combination of forces—the efforts of the countries themselves, the development initiatives undertaken by private multinational firms, the work of the great foundations and of the International Research Network, loans of the World Bank, the multilateral efforts of the Food and Agricultural Organization (FAO) of the United Nations and other agencies, such as the private voluntary organizations (PVOs), plus technical assistance through US-AID—these together have improved the state of the art in agriculture fast enough to keep ahead of the population explosion. This is a great achievement, a sustained rate of increase worldwide in agricultural production that previously had never been achieved. The contributions of the various initiatives cannot be separately identified. Most of the credit undoubtedly belongs to the developing countries themselves. Unfortunately, Africa is the exception to this favorable report. The per-capita output there is falling for reasons not wholly agreed upon.

The American Agricultural Revolution of the 20th century is being exported; adapted rather than adopted. It can't simply be transplanted. It is being transformed. In the Third World it is recognizable, reminiscent of its source and a worthy offspring of its parent.

Section II

The Deeper Dimension

Chapter 9

BIRTH PANGS— AND A LIVE BIRTH

How did farmers cope with the Agricultural Revolution of the 20th century? If we are talking about survivors, the answer is "Amazingly well." In 1910 the average income per person on farms earned from farming was 71 percent *less* than the average income per person not on farms (USDA, *Agricultural Statistics*, 1942). (No wonder there was off-farm migration.) In 1997 the average farm operator's household income was 5 percent *greater* than the United States average household income (USDA, *Agricultural Statistics*, 1999). Of the incomes of people classified by the census as farmers, 84 percent is from nonfarm sources.

INCOME AND NET WORTH

Income is like the flow of a pipe into a stock tank. What of net worth, which is the accumulated flow of the pipe into the stock tank, minus evaporation and minus what the cattle drink? Both are important; one the inflow (income) and the other the reserve (net worth). Bonnen and Schweikhardt (1998) report that average farm family wealth is now greater than that of the average U.S. family. Gardner (1992) confirms this finding. A Federal Reserve Board survey of 1962 found the mean net worth of farm operators to be $44,000, about three times the net worth of nonfarm occupations (Projector, 1964). In 1996 the total farm real estate and non-real estate debt was $166 billion. On January 1, 1997, the total value of land and buildings was $910 billion, giving farmers overall equity of 82 percent, unequally distributed (USDA, *Agricultural Statistics*, 1998). It seems reasonable to think that, at a time of economic difficulty, farmers might draw on their reserve and increase their debt to weather a storm. But the uneven nature of the debt burden, farm by farm, makes this a difficult remedy for the farmers in deepest trouble.

The amenities of rural life have advanced dramatically during the 20th century. In 1900 farm families had almost no electricity, cars, or phones. Now almost every farm has these conveniences in addition to the new blessings the century has brought: central heating, radio, television, and the computer. At the start of the century, most farm families lived on dirt roads. Now most farmers are out of the mud; rural roads generally are hard-surfaced. Health service has improved, lengthening the life span. At the beginning of the 20th century, high school was a rarity for farm youth. By 1995, more than 80 percent of farm youth had attended high school and 36 percent had studied beyond that level.

That these advances could have come about despite the Great Depression and the five wars of the 20th century is truly remarkable. Year by year the changes were almost imperceptible; cumulatively they are astonishing.

NOT A FACELESS GROUP

Thus far we have been treating farm people as a whole. But we get limited overall relevance by grouping together the Kentucky grower who has two acres of tobacco, the Kansas farmer who has 2,000 acres of wheat, and the Ohio part-time farmer who gets most of his income from assembling automobiles. However, all of these are officially defined as farmers by the U.S. Department of Agriculture (USDA) because they produce, for sale, $1,000 or more of farm products annually. This figure is ridiculously low, made so deliberately because of the political interest to report as large a number of farmers as possible. Statistically, marketing a few cows or selling a few acres of corn makes a person a farmer. Jim Bonnen, of Michigan State University, says he could raise $1,000 worth of asparagus in his backyard.

Bonnen and Schweikhardt divide agriculture into three groups that might be called the winners, the contenders, and the losers. In addition, there are those we might call the agricultural alumni.

THE WINNERS

This is the commercial agricultural sector, which is made up of an economically viable commercial farming component, a business sector providing farm inputs, and another business sector that processes and markets farm products to the consumer. Note that Bonnen and Schweikhardt integrate the three elements of the food system: providers of input items, production agriculture, and marketing of the product. This is a modern, enlightened concept of agriculture, but it is not in keeping with traditional views of farm people. Farmers persist in thinking that agribusiness and consumers are their adversaries.

The economically viable farm component of agriculture may be said to consist of farms with commercial sales in excess of $250,000 each. Over the years, these large farmers have increasingly dominated agricultural production. In 1989 they constituted 5.9 percent of all farms and had an average net worth of $1,331,000 (Table 9-1). They were, Gardner (1992) said, a class of millionaires with net incomes placing them in the upper 1 percent of all U.S. households. They are the survivors of the Agricultural Revolution. This group embodies the chief constituents of the farm politicians and the chief supporters of the farm lobby. The major share of the political

TABLE 9-1. Characteristics of U.S. farms by economic class, 1989.

	Annual Sales		
	$250,000 or more	$40,000–$249,999	Less than $40,000
Share of all farms	5.9%	24.8%	69.4%
Farming primary occupation	96.7%	90.9%	39.6%
	$ Per Farm		
Net cash income from farming	252,000	38,000	200
Off-farm income	21,000	18,000	30,000
Total income	273,000	56,000	30,200
Assets	1,600,000	610,000	230,000
Debt	269,000	80,000	18,000
Net worth	1,331,000	530,000	212,000

Source: Taken from Gardner, 1992, with permission.

controversy in agricultural circles emanates from this group. These people have legitimate concerns: unstable prices, gyrating export markets, variable weather, galloping technology, regulatory edicts, and decisions regarding government legislation. As a group they tend to favor expanding export markets and are quick to adopt new technology. They resent regulatory edicts. They are risk takers who sometimes win and sometimes fail. But many of them still prefer that risks be socialized.

Government payments on supported commodities accounted for more than 20 percent of *total* net farm income in the late 1960s and an even higher percentage throughout much of the 1980s. Under the 1985 farm bill, 15 percent of all these payments were received by less than 2 percent of U.S. farms with annual gross sales of more than $500,000. At the other extreme, nearly half of all U.S. farms under financial distress during this period received no direct payments at all (Orden et al., 1999). The chief concerns of this wealthy group are to keep ahead of the Agricultural Revolution and to deal with the environmental issues produced thereby. As if they were on a treadmill, these farmers struggle to stay ahead of the times.

THE CONTENDERS

This midsized class accounts for one fourth of all farms, with sales between $40,000 and $250,000. With an average net income—farm and off-farm—of $56,000, this group can hardly be said to be economically distressed, though some of them are indeed in distress. They received, on average, a dollar from off-farm sources for every two dollars earned in agriculture. These farmers are often on a knife-edge between growing into a winner and becoming a loser. While they may not receive the bulk of government support offered to agriculture, the support they do receive constitutes a relatively larger share of their income. What happens to these farmers is a critical policy issue as concentration of wealth in rural areas becomes more severe.

THE LOSERS

The smallest group, farmers with annual sales of less than $40,000, constituted 69 percent of all farms. On average, they netted only $200 from farming yet had an average of $30,000 per year from off-farm income. While many in this group were abjectly poor, as a whole they could hardly be so characterized. These were mostly part-time farmers. Most of them lost money on their farming operations. This group totaled nearly 70 percent of all farms and produced only a minor share of what goes to market. These were the people who, when the Agricultural Revolution came, stayed small and opted out unless they found an off-farm job to supplement their small farm enterprise and so survived. They are claimed as constituents by the politicians, but rarely are they recipients of political favors. They are a group with a very wide range of incomes, the overwhelming proportion of it from off-farm sources.

The wish to believe that farm people are poor, which was true early in the 20th century and is the persistent mantra of many urban people and many farm politicians, is an idea that dies hard. The early disparity was corrected by the Agricultural Revolution at great cost to many who were vulnerable at the time it struck.

THE AGRICULTURAL ALUMNI

These people said good-bye to agriculture when the Agricultural Revolution struck. There were 29 million farm people in 1900; by the year 2000 it is estimated that there will be fewer than 5 million, a fearful rate of attrition. Lenin said, "You can't make an omelet without breaking eggs." The Agricultural Revolution broke a vast number of eggs to get the omelet we call modern agriculture. Most of the scientists who produced the Agricultural Revolution did not know they were breaking eggs or, if they knew, chose to ignore that fact.

The sympathies of these former farmers and their progeny are with agriculture, which is their heritage. They watch from the stands and applaud when the home team wins. They provide critical support for agricultural policy and institutions. They are a connection between urban and rural America. They are not numbered among the 5 million farm people reported by the USDA. Those with education and zeal, who live in areas of substantial off-farm opportunities, have done well. Former farmers are not counted in reports of agriculture's per-capita income or net worth; they are outside of agriculture. They have been graduated; call them the agricultural alumni.

INDUSTRY'S INFLUENCE

The United States was most fortunate in that, when the Agricultural Revolution came along, industry was expanding, providing employment opportunities for dispossessed farmers. By offering attractive employment opportunities, industrial growth created a need to adopt the technology of the Agricultural Revolution. Despite social problems associated with labor's leaving agriculture, on balance the United States absorbed that labor in nonagricultural employment, and most of those who were relocated fared well.

Some Third World countries have little industrial growth and therefore limited opportunity to absorb displaced farmers. Agricultural-development workers in Third World countries are aware of this constraint and are properly hesitant to push agricultural initiatives that would displace farm people. Similar problems are occurring in economies undergoing transition from central planning to market economies. Large agricultural and industrial enterprises are shedding labor rapidly. This labor becomes unemployed or adopts subsistence farming. American agriculture, with its increasing off-farm opportunities, avoided the dilemma.

Child of the 20th Century

The Agricultural Revolution is a child of the 20th century; that much is clear. The date of birth is clear. The exact date or dates of conception are uncertain. Perhaps there was multiple parentage. There were birth pangs—displaced people, altered hopes, shattered dreams. But it was a live birth, and the child is a lusty one, full of promise and clearly destined to live and grow during the centuries ahead. Whatever there be—pain at the pangs of birth or joy at the appearance of the new child, the words of Omar Khayyam are relevant:

> The Moving Finger writes; and, having writ,
> Moves on: Nor all your Piety nor Wit
> Shall lure it back to cancel half a Line,
> Nor all your Tears wash out a Word of it.

Chapter 10

CONSUMER BENEFITS

The American consumer buys food—the best and safest food of any country on the globe—for the lowest percentage of family income than is true of any of the 64 countries for which statistics are available (Table 10-1). While this is in large measure a consequence of the Agricultural Revolution, it is also a result of the increased incomes of American consumers.

Visitors to the United States are amazed at the quantity, variety, and quality of food items in American grocery stores. The typical supermarket carries 16,000 items with 35,000 square feet of floor space.

Americans use—and misuse—food in a lavish manner. Restaurants and hotels serve larger portions than is appropriate and much food is left on the plate. The U.S. Department of Agriculture (USDA) reports that 15 percent of the average school lunch meal ends up in the garbage can.

The real price of food (actual price corrected for inflation) has been remarkably stable for many years. This is the result of two offsetting trends. On one hand, the farm price of food has been falling. The ratio of prices received by farmers to prices paid by farmers, which stood at 100 in 1910–14), fell to a ratio of 47 in 1996 (USDA,

TABLE 10-1. Budget share of income spent for food in 64 countries (in order of lowest to highest percent, 1985).

Country	% for Food	Country	% for Food	Country	% for Food
United States	10	Spain	24	Morocco	39
Canada	11	Cameroon	5	Saint Lucia	39
West Germany	11	Botswana	26	Suriname	39
Hong Kong	12	Hungary	26	Zimbabwe	39
United Kingdom	12	Yugoslavia	27	Ivory Coast	40
Australia	13	Greece	30	Turkey	40
New Zealand	13	Malawi	30	Congo	42
Bahamas	14	Rwanda	30	Sri Lanka	43
Denmark	14	Swaziland	30	Nigeria	48
Netherlands	14	Thailand	30	Egypt	50
Luxembourg	15	Poland	31	Ethiopia	50
Norway	15	Grenada	33	Senegal	50
Sweden	15	Barbados	34	Bangladesh	51
Austria	16	Portugal	34	India	52
Belgium	16	Korea	35	Philippines	52
Finland	17	Benin	36	Sierra Leone	56
France	17	Jamaica	36	Nepal	57
Japan	17	Zambia	36	Mali	58
Italy	19	Pakistan	37	Madagascar	59
Trinidad	19	Tunisia	37	Tanzania	62
Ireland	24	Iran	38		
Mauritania	24	Kenya	39		

Source: United Nations, 1992

Agricultural Statistics, 1998). This is the result of the cost-lowering forces of the Agricultural Revolution.

On the other hand, more marketing services have been added, counteracting the lower cost of food on the farm level. In 1954 food services added to the raw product of the farm took 26 percent of the consumer's food dollar. By 1986 this percentage had risen to 44. The following were increased:

- Eating out
- Convenience foods
- Packaging, advertising, and promotion
- Transportation (with the shift from in-season, locally produced food to food supplied from distant sources)

These increases are evidence of greater affluence made possible in part by the Agricultural Revolution and in part by greater efficiency in the nonagricultural sector. The phenomenal growth of the nonagricultural sector during the 20th century eased the pain of what would otherwise have been an agonizing experience for American farmers. The Agricultural Revolution, which reduced the real cost of food at the farm level, made possible a stable real cost of food to the consumer along with the many added services the marketing system provided.

American lifestyle has experienced a great transformation. Many families no longer sit down together at meals. Delicatessen departments in supermarkets, fast-food outlets, and the microwave oven are the result of a social and economic revolution that paralleled and was accommodated by the Agricultural Revolution.

Farmers complain that they receive a declining share of the consumer's dollar, which indeed they do. In the early days, when farmers sold much of their food directly to consumers, they received nearly 100 percent of the consumer's food dollar. In 1996, when all the food services had been added, they received only 29 percent. But they did better than when they received nearly all of it. Farmers who sell "u-pick" blueberries receive 100 percent of the consumer's dollar, but they do no better than the wheat farmer, who receives only 7 percent. The percentage of the food dollar received by farmers is a poor criterion for farmer well-being.

Many farmers and consumers think that food processors and food chains make exorbitant profits. The record shows that the food business is very competitive and that profits are generally in line with returns from other investments.

When there is a glut of some food, for example pork, grocers are slow in reducing their prices. They know that consumers dislike fluctuating prices. They know that if they drop the price at a time of abundance, they will have to increase it at a later time of scarcity. They know that they get more complaints when they raise the price than applause when they reduce it. It would be well economically if they reduced the price at a time of abundance and so moved more produce, relieving the overload and strengthening the weak price at the farm level. But all their instincts are against it. They tend to pool the system instead of particularizing it item by item. What they make now on meat they may lose on produce, and at a later time this may be reversed.

Contrary to what many farmers and consumers think, the food system is consumer driven as well as supply driven. Food moves from the farm to the consumer, but the price-making forces move in the opposite direction. Of course, advertising, promotion, and food fads affect the consumer's desires. But whatever those desires are and whatever their cause, they are reflected in what the consumer will pay for the food he or she wants. And this reflects back to the wholesaler, the food processor, and ultimately the farmer. It was in accordance with these signals that farmers shifted from the producer of fat to lean hogs and from Guernsey to Holstein cattle.

During the world food scare of the 1970s, there was concern that the United States might run out of food. The USDA was sharply criticized for permitting the drawdown of our food stocks. The scare was ill-founded. We have an enormous food reserve in the form of livestock on the hoof. At a time of need, we can consume our livestock and then eat the grain that the livestock would otherwise have eaten. We can thus increase our food supply by several magnitudes. With this flexibility, we met the drought problem of the 1930s and 1970s.

There is much concern about the conversion of farmland to nonfarm uses, and for good reasons: (1) the desire for open space, (2) the wish for a reasonable balance between rural and urban cultures, and (3) nostalgia for things that once were. But an alleged threat to our potential food production is not a valid reason. By government policy we hold land out of production, in part to conserve it for future generations and in part to shorten the supply of food and so enhance the price. In 1987 the government paid farmers to hold over 60 million acres out of production. In fiscal 1998, the government spent $1.8 billion on a conservation reserve program, restoring land declared to be environmentally sensitive. Some 300 million acres are classed as prime farmland. The Soil Conservation Service says there are approximately 100 million acres not now tilled that could be converted to cropland within two decades if prices were attractive. Admittedly, these acres would be less productive and more erosive than those now in use. The area of cropland shrinks by about 0.4 percent per year (Cotner, 1976).

The real price of food at the farm level has been on a long-time downward trend, an indication of abundance rather than of scarcity. When the authors sit down to do their serious worrying, they do not worry about the United States running out of food or farmland. What we do worry about is the disparity of wealth and income between the rich and the poor in the United States and worldwide, the wastage of agricultural resources through wind and water erosion, and unrestrained population growth in the developing world.

During the 1960s, when the senior author was in the USDA, he called the statistician in the then U.S. Department of Health, Education, and Welfare (now the Department of Health and Human Services) to learn about human maladies such as scurvy, rickets, beri-beri, and goiter resulting from dietary deficiencies. He was told that these afflictions had become so few that they were no longer monitored. This reflects the increase in abundant, nutritious food as a consequence of the Agricultural Revolution and the advance of income.

For poor people who have difficulty paying the going price for food, the USDA has a Food Stamp program. In 1996, 26 million people (10 percent of the population) were on food stamps. Since then the number has come down sharply.

Scares about contaminated food make headlines, precisely because they are so few and hence newsworthy. There are an estimated 92 million food-related illnesses each year, with 9,000 deaths. Yet roughly 290 billion meals are consumed each year. That gives a person a 0.03 percent chance of getting sick from a given meal, or one illness in three years. Ever since 1906, as a result of Upton Sinclair's muck-raking book *The Jungle,* the USDA has had a food inspection service to ensure the wholesomeness of the American food supply.

One tends to think of the Agricultural Revolution as affecting only things agricultural—crops and livestock, but it also affects the human species. Better nutrition, health care, and sanitation also impact people. Consider the consequences:

- We live longer and are healthier than previous generations. Since 1900, 29 years have been added to the human life span. In 1900, the average life span was 47 years. During 1990–95, it was 76 years.
- Athletic records are continually being broken.
- The younger generation exceeds its elders in stature.

All these changes must be related, at least in part, to better nutrition and so to the Agricultural Revolution. Interestingly, the study of human nutrition lagged behind the study of plant and animal nutrition launched by Justus von Liebig (1803–1873). Farmers had economic incentives to improve nutrition of plants and animals; the need to improve human nutrition was more obscure.

Farmers number about 5 million. Together with their input suppliers, they feed an American population of about 270 million plus many millions overseas. To quote Winston Churchill's tribute to the Royal Air Force, which saved England from the Nazi invasion: "Never before have so many owed so much to so few." The marvel of the American food system is not shared by many other countries.

WORLD HUNGER

The Agricultural Revolution provided astounding increases in the production of many crops, particularly wheat, corn, and rice, the backbone of the world food supply. To the affluent American consumer who spends only a limited share of income on food (variously estimated at 10 to 14 percent depending in part on whether beverages are included), this means an increase in the supply and quality of food at a reduced share of income. This is a substantial but hardly life-altering gain. But to many consumers in the poverty-stricken Third World, where large numbers of people live on the borderline of hunger and spend as much as 50 percent of their incomes on food, the Agricultural Revolution can mean the difference between life and death. The World Bank estimates that 800 million people are malnourished. Table 10-1 reveals that countries that have good agricultural endowment and are industrialized have done well in meeting food needs.

All three of the great cereals—corn, wheat, and rice—were transformed during the Agricultural Revolution. Corn was hybridized early in the century by Shull, East, Jones, and Wallace. Wheat was improved dramatically in midcentury by the American scientist Norman Borlaug, who worked in Mexico. Rice followed the path of wheat blazed by Borlaug; in 1966 "miracle rice" was released by the International Rice Research Institute in the Philippines. All three of these genetically transformed crops were supported by advances in fertilization, pest control, and improved tillage—likewise products of the Agricultural Revolution. Rice yields per acre in India have increased by 80 percent and wheat yields by 50 percent since the middle 1960s.

In primitive society, before the advent of sanitation and disease control, human birth and death rates were in rough adjustment at fearfully high levels. Sanitation, inoculation, and pest control

TABLE 10-2. Human death rate in Taiwan.

Year	No. per 1,000
Before 1943	30 ± 5
1943	20
1947	18
1956	8
1983	5

lengthened human life, just as they improved crop yields. So the human death rate in the Third World fell. In Taiwan, the death rate began falling about 1920. The decline was one of most rapid on record (Table 10-2).

With the birth rate continuing to be historically high and the death rate reduced, countries are experiencing what is called the *demographic transition,* during which population numbers increase dramatically. Typically, a generation (or two or three generations) after the death rate begins to fall, the birth rate also begins to fall. In time, both reach approximate equilibrium at a vastly increased number of people. This has already happened in the developed countries. During the demographic transition, the Third World is experiencing the population explosion so vividly described by Ehrlich, the Paddock brothers, Garret Hardin, Donella Meadows (author of the *Report to the Club of Rome*), and by Lester Brown of World Watch (Figs 10–1 and 10–2).

Had it not been for the Green Revolution, Malthusian-style mass starvation would have occurred not in the United States but in the Third World. The mistake made by the above-named prophets of disaster is that they failed to take into account the growth factor of the Agricultural Revolution, which was worldwide.

The world hunger problem seems likely to recede for generations following our own. In the long run, well into the 21st century, economic development, family planning, and education seem likely to reduce the birth rate and move the demographic transition toward completion. Estimates by the United Nations of the prospective world population have come down partly because of the reduced birth rates and partly because of the spread of AIDS in Africa. In October 1998, the United Nations Population Division projected (medium estimate) that by the year 2050, the world

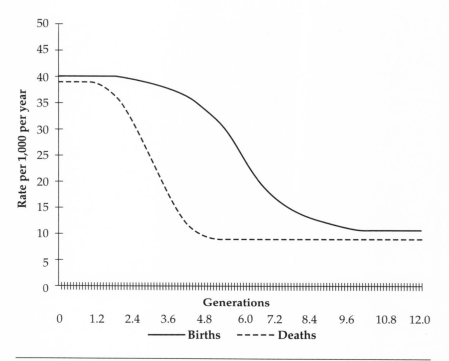

FIGURE 10-1. The demographic transition (generalized), in large part, in the United States and Western Europe.

population would total 8.9 billion, compared with a 1996 estimate of 9.4 billion (Population Action International, 1998). This projected number would be approximately twice the level of 1940.

The world food problem is not a lack of food; it is really an income problem. The overall supply of food is adequate, but poor people lack the income with which to buy it. Even in developed nations like the United States some people are hungry because they lack the money to purchase food. Hunger is more a distribution problem than a problem of total supply.

With the passage of time, the growing population has laid increasing demands on land and water resources. Cultivation has pushed onto fragile lands, inviting erosion. Forest lands are denuded for firewood and tillage, leaving watersheds unprotected. People crowd together in enormous cities; civil disturbance associated with excessive numbers becomes a threat. Resources are depleted, raising a question as to whether science and technology can develop needed substitutes.

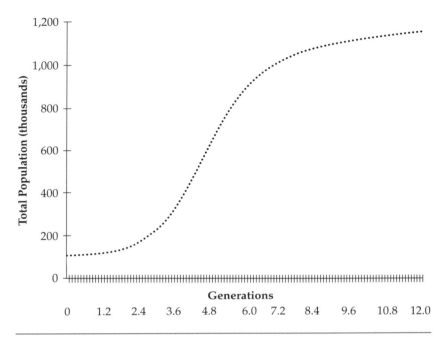

FIGURE 10-2. The population explosion (generalized). Idealized chart for the population experience of a Third World developing country.

If population continues to grow, and if these trends continue unabated, grave problems may arise. It is indeed possible that if food problems were solved, unrestrained population growth might plunge us into other difficulties. The Rev. Thomas Robert Malthus, writing in 1798, identified the food supply as the ultimate limitation on population growth. He may have been in error. Rampant population growth threatens not just the food supply but also the resource base and domestic tranquility. These are forces with which the Agricultural Revolution cannot cope; it can only relax the land constraint imposed on food production.

The Agricultural Revolution of the 20th century was not just American and not just agricultural. It created both great benefits and real problems. It was a revolution not of swords and guns but of test tubes and microscopes. It was life saving rather than life destroying. What a privilege to have witnessed and participated in this profound change, and what a challenge now to record it!

Chapter 11

Political Issues Arising From The Agricultural Revolution

During the first years of the 20th century, political issues in agriculture centered around monetary policy, tariffs, and restraint of monopoly. During the '20s, issues shifted to two-price plans and the effort to lift the domestic price above the world level. With the '30s came the attempt to increase agricultural prices by restricting production. With occasional digressions, price supports and production controls held stage center for more than 60 years.

The story is told that in the early days, when farm policy strategy was being developed, Ed O'Neil, prominent architect of farm policy, consulted with John L. Lewis, the labor leader, about farm-policy strategy. "If I were you," Lewis said, "I'd go for a higher price. We've done great things for labor—better working conditions, probably most important. But the workers don't understand that as well as they do a higher wage. So we go for higher wages. I would think that for farmers a higher price would have the same appeal."

For this reason and some others, higher prices, calibrated by parity, became the mantra of farm policy for 60 years.

Farmers have historically measured the prices they receive by the parity ratio: an index of the prices they receive relative to an index of the prices they pay, with the base during 1910–14 equal to 100. This ratio has drifted downward throughout the 20th century (Table 11-1). This downward drift seems to give credence to the farmer's belief that the economic scales are tilted against him. Actually, the decline is the consequence of the cost-lowering effect of the Agricultural Revolution. The supply-increasing impacts of the Agricultural Revolution on a price-insensitive demand triggered much of the farm policy agenda. Price supports, land retirements, and demand promotion policies were attempts to counteract the effects of the Agricultural Revolution. The productivity gains of the Agricultural Revolution are reflected in a lower price—one of the laws of economics discussed in Chapter 6.

Learn now what is probably the most persuasive evidence of the Agricultural Revolution. At the end of 1998, average per-capita incomes of farm people were virtually the same as average per-capita incomes of nonfarm people, and the mean net worth of farm families was roughly four times as great as that of nonfarm families (and this with the parity ratio at 41 percent of its 1910–14 base). All

TABLE 11-1. Parity ratio in U.S. agriculture (five-year average).

Dates	Parity Ratio 1910–14 = 100
1910–1914	100
1915–1919	109
1920–1924	89
1925–1929	91
1930–1934	69
1935–1939	86
1940–1944	100
1945–1949	109
1950–1954	98
1955–1959	83
1960–1964	79
1965–1969	76
1970–1974	78
1975–1979	71
1980–1984	59
1985–1989	52
1990–1994	48

Source: USDA, *Agricultural Statistics,* various issues.

of this occurred because the Agricultural Revolution boosted yields, replaced expensive inputs, captured economics of scale, and increased farm size. Historic price-support and production-control programs got into trouble. In 1996 Congress enacted what became known as the Freedom to Farm Act. It decoupled production control from benefit payments (Schertz and Doering, 1999). A series of events followed that cast doubts on the wisdom of Freedom to Farm: big crops; large carryover; reduced exports; and falling prices for corn, wheat, soybeans, and hogs. The government voted for large sums of money to aid farmers, apart from production control or price support.

There was disappointment with the Freedom to Farm Act but confusion about what might supplant it, and there was general unwillingness to return to the old legislation. Orden et al. (1999) considered that the old price-support and production-control system was unsatisfactory. The most promising way to extricate ourselves from it, they said, was to cash it out; in other words, pay farmers to relinquish their acreage allotments and price supports.

Throughout the 20th century, the government sought to hold the structure of agriculture, as nearly as possible, in the pattern of the past. This effort was made impossible by the fact that farming was in the midst of an agricultural revolution.

A flawed diagnosis produced a faulty prescription.

Agricultural policy historically was based on the idea of the family farm, the competitive society, and self-interest on the part of the farmer. The prevailing farm motivation was "If it will make money, do it." There was little regard on the farm or in Congress for side effects or for what the economist calls "externalities."

Balancing the social cost (disadvantages imposed on the society in general) with private gain was made with minimal concern for the overall well-being. Now, at the end of the 20th century and the beginning of the 21st, there is increased concern for the public interest. Economists would say that marginal private gain and marginal public costs are being equated at a new level. This was largely the result of a change in what is called the social welfare function. Many in positions of power are little aware of the change or are reluctant to acknowledge it.

Technology is the engine of change. Some time after technical change occurs comes a modification, often reluctantly made, of institutions such as educational establishments and of the mission of farm organizations. But political rhetoric continues to repeat the old concepts. Finally, change is acknowledged by all. But by that time new issues have emerged and the process is repeated. So we have a built-in lag in policy making.

These stages of change are discernible as we review how agriculture moved from the 19th to the 20th century. Mechanical and electrical power began transforming agriculture early in the century, and an industrialized agriculture was in gestation. It took the land grant colleges some time to realize this change, and it took the politicians even longer. Politically, we continue to debate old issues of declining relevance.

Late in the 20th century, as the nonfarm population became sensitive to the side effects of modern agriculture, farmers began losing control of the farm policy agenda. A chief reason for wanting control is that such control makes it possible to suppress undesired issues. By the 1970s, '80s, and '90s, these undesired issues could no longer be avoided. To the distress of farmers, a number of them found their way into the public forum. Many of the new issues were, directly or indirectly, a consequence of the Agricultural Revolution, all representing situations in which the social costs were not in balance with private gains. Some of the following developments now have antifarm advocates:

- Heavy fertilization caused groundwater pollution, endangering human health, and led to the eutrophication of ponds and streams.
- Chemical pest control threatened to eliminate some desired species of animal life.
- Industrialized agriculture, a product of the Agricultural Revolution, was seen as a threat to the family farm, which it is.
- Food products grown on heavily fertilized fields were said to be nutritionally deficient, which is factually doubtful but politically potent.
- Low-input agriculture was advocated, natural foods were pushed, and "organically grown" became a desired label. "Food additive" acquired a negative connotation. "Food safety" took on added importance.
- Advocates of animal rights, offended by the new practice of confinement housing and feeding, injected an ethical dimension sometimes real and sometimes imagined into farm policy where it had not previously existed.

- Genetically modified organisms resulted in confrontation between farmer and consumer groups.
- An antiscience, antitechnology attitude arose, challenging the basis of the Agricultural Revolution.
- Weapons of mass destruction such as sarin, anthrax, and the atomic bomb entered the arsenals of the United States and other countries, deepening the distrust of science.
- The cloning of Dolly, the Scottish sheep, aroused anxiety about the possibility of cloning the human species. Korean researchers claim to have cloned human tissue.
- Farm laborers contended that their health was endangered by having to work in fields newly sprayed with pesticides.
- The use of fuel in modern agriculture was said to add carbon dioxide to the atmosphere and thus contribute to the problem of global warming.
- Large-scale confinement feeding, particularly of hogs, came under heavy attack from neighbors and neighboring towns. The old farm axiom, "It smells like money to me," was no longer persuasive to the public.
- Farmers with new drainage equipment sought to dry out their ponds to grow corn but were restrained by wetland preservation.
- The general feeding of antibiotics to farm animals, it was feared, might cause transformed organisms to escape to the human population, where they would be immune to the standard antibiotics.
- The Agricultural Revolution had resulted in the consolidation of farm units so that farm ownership had become concentrated, contrary to Jefferson's ideal of the yeoman farmer and broad-based farm ownership.
- The widening gap in agriculture between the wealthy and the poor was offensive to many.

Of course, while new issues emerged, the old issues of foreign trade, price supports, and production controls did not go away. But they now had to share time, concern, and money with the new issues, which arose to challenge an agricultural agenda that had been almost exclusively in the hands of the farm lobby. The farm lobby at first adopted techniques of denial, then of hostility and resistance.

The new issues ignored the beneficial results of the Agricultural Revolution (documented in the previous chapter), which farmers continued to declare, with diminishing effect. Perhaps affluent consumers have become less concerned about the availability of their food supply and have shifted their awareness to what they consider its "quality" and its associated institutions. These issues are put forward by well-financed lobbying groups, often with more zeal than objective fact.

The farmers who previously had no need to deal with these issues lacked experience in coping with them. The first reaction was to deny the validity of the protests. When that failed, there was an effort at cooptation, the present strategy.

New Initiatives

The U.S. Department of Agriculture (USDA), aware of the shift in the public mood, supported a number of initiatives, some old and some new, in part to build a backfire against the environmental concerns that had entered the farm policy agenda. These initiatives had to do mostly with better resource use. They were adopted in part because they recognized the legitimacy of many of these concerns (USDA/ERS, 1998).

Establish Accountability

As the century drew toward its close, there was a fumbling effort to establish a new ethic: accountability—the taking of responsibility for an action, be it good or bad.

But, more questions arose: Accountability to whom? Accountability by what criterion? What combination of education, incentives, and coercion is needed to achieve desired objectives?

Clearly, the Agricultural Revolution injected new items into the agricultural policy agenda. It increased supplies relative to available markets, which resulted in lowered prices that were not acceptable to farmers and brought governmentally enforced acreage reductions. The revolution also produced the externalities previously mentioned. These externalities once were considered to be of little consequence. Now, however, society places a high value on what has come to be called "the new agenda," which replaces the old agenda of price supports and production controls. The Agricultural Revolution provides a new focus for an issue that has been with us from the beginning: What degree of individual freedom and public involvement is needed in decision making? And if the decision making is centralized, who makes the decisions, and in whose behalf are they made?

We see clearly that the Agricultural Revolution is technological. Less visible is the fact that it is also social, and more obscure is its political impact.

Section III

Looking Ahead

Chapter 12

Into the 21st Century

The most significant agricultural breakthrough of the 20th century, it seems to us, was the discovery in 1953 by Watson and Crick of deoxyribonucleic acid (DNA), the "secret of life" and the ensuing manipulation variously called "genetic engineering," "gene-splicing," and "recombinant DNA." These initiatives made possible the modification of existing species of plants and animals and even, incredibly, the creation of something like new species. The profound significance of the new capability is slowly entering public awareness. There is a struggle now to substitute some kind of accountability for what was formerly the blank check given to science.

A hundred years is long enough for a new age to emerge. Will technology continue to be our major preoccupation during the 21st century? Will materialism be in some degree replaced by concern for the aesthetic? For many years we moved away from the individual and toward group action. Will this continue, or has this trend run its course? Will the present institutional arrangement continue for the farm and nonfarm sectors, or will there be change? If so, what?

Science and technology will no doubt continue in the saddle for many years to come. It is said that more scientists are now living than have lived and died in all the years up to now. The American educational system is tilted toward mathematics, science, and commercial skills and away from the humanities. That scientists will continue with research and discovery is incontestable; the foundation for a scientific agriculture has been laid.

But the previous unquestioned acceptance of science and technology as the benefactors of the human race is being challenged. Science and technology will increasingly be asked to show broad-scale benefits to the human race. The wonders and marvels of new knowledge and methods, formerly a sufficient justification, will not suffice. We may make improved use of our technological competence in the years ahead and strike a better balance between things material and things of the heart, mind, and spirit. A hundred years from now the gross domestic product may not be the decisive public-policy statistic; the gross national product may be adjusted for externalities as is done in Holland.

Economics as a discipline was developed late in the 18th century when the prevailing activity was agriculture, the dominant focus was on the individual, and the assumed motive was self-interest. Then followed the Industrial Revolution of the 19th century and the Agricultural Revolution of the 20th century, modifying the setting for the discipline. Institutional economics, which admits ethics, law, politics, and social subjects into the decision theater, struggled to gain acceptance during the 20th century and may achieve recognition during the 21st.

But science and technology will still be important. What changes might we see? (Paarlberg, 1976)

The plow, which was the symbol of agriculture during much of the 20th century and has been the emblem of the U.S. Department of Agriculture, will gradually be retired. We will rediscover the wisdom of Native Americans who, the first time they saw a plowed field, commented, "Wrong side up." The wind and water erosion caused by the plow is incalculable. We are learning better how to grow crops using minimum tillage. We will use various kinds of mulches. What is left of our soil we will try to keep in place rather than sending it flying through the air and drifting down the river.

What technological breakthroughs might occur in agriculture during the 21st century? One can do no more than surmise that some of the hoped-for advances will in fact occur, such as the following:

- Advances in use of the microchip that today defy the imagination.
- Upgrading the protein content of the cereal grains and other crops by biotechnology.
- Soil-management techniques that would permit the use for agriculture of the fragile soils of the tropical rain forest. (We may learn to grow trees for oil or fruit or wood, protecting the soil from the scorching sun and the pounding rain, supplying the oxygen we need to balance the gasses of the atmosphere.)
- More biological control of harmful insects and diseases.
- Control of the tsetse fly, the vector of sleeping sickness in Africa, thereby opening for agricultural use vast areas of that continent that are not now in agricultural production.
- Improved long-range weather prediction technology.
- More use of satellites for worldwide crop reporting and global positioning.
- Use of biotechnology in fixing nitrogen for new groups of plants, thus cutting down the need for commercial fertilizer.
- Production of grain without the need for male plant parts, by the process of apomixis.
- Desalination of seawater and use of biotechnology to produce salt-tolerant crops, thus permitting human habitation and agricultural production in lands now unused.
- Acceleration of hydroponics, which has long been latent.
- Conquest of the fuel problem by fuel cells, safe use of nuclear energy, and use of ethanol produced by biotechnology from plant biomass.
- More intelligent environmental control in response to the public interest. (The problem of groundwater pollution, which has received inadequate attention, will be addressed in earnest.)
- Advances in food technology that would allow the provision of meat analogs to the many millions who cannot afford palatable and nutritious meat, milk, and eggs.
- Widespread sterilization and radiation of food so as to permit long shelf life without refrigeration.
- Institutional changes that will permit us to steer more responsibly the industrialized agricultural juggernaut we have created.

- Use of microbial action on various feedstocks for the direct production for feed and food.
- Advances in technology that will allow better understanding of global warming and of el Niño and la Niña.

Other changes might occur also, such as the following:

- Land grant colleges will be modified in the direction of greater emphasis on basic research and blurring of department boundaries; schools of agriculture will become more integrated with other schools of the university.
- Intellectual property rights will be reinterpreted to reflect adequate concern for nonintellectuals.
- Revision of government programs for sugar, peanuts, and tobacco will more adequately reflect the public interest.
- Systems for distribution and educational advances will be improved so as to minimize the twin problems of overeating and poverty-related malnutrition.
- Use of computers will increase. During the 21st century, the person who cannot use the computer will be in the same position as the person of the 20th century who could not use the automobile.
- The social status of the farm family will advance.
- (Most important of all) Family planning methods will improve and public acceptance of the replacement-sized family will increase, so that the human species might move out from under the Malthusian shadow.

WHICH WILL SURVIVE?

Of course, how many of these technological advances and other changes might occur we cannot now know. We can only ask that if this book should be preserved and read a hundred years from now, the reader would be a charitable and forgiving person.

Of one thing we can be sure: whatever the technical developments in agriculture may be, its poetry will continue. The wonders of life, growth and death, the cycle of the seasons, the marvels of continuous creation so clearly manifested in agriculture—felt but not expressed by the person on the land—will still be with us. In fact, it might be said that these are the most enduring features of agriculture and that the subject matter of the poet permits him or her to be the best, though the least understood, prognosticator of all.

What about the institutional arrangements within which crops and livestock will be produced? For one thing, agriculture will continue to lose its uniqueness. Farm production will be merged in an integrated fashion with the acquisition of input items and with the processing, transporting, financing, merchandising, and consumption of the product. There will then no longer be a Department of Agriculture; there will be a Ministry of Food, as has already occurred in a number of other countries.

A mixed farming system will emerge. There will be large-scale, integrated units. The production of crops and livestock will require farms so large—so much land, capital, and specialized managerial skill—that one person will be unlikely to supply them all. In commercial agriculture, the nearest thing to the family farmer will be a farm operator who lives on the land with his or her family, rents the farm, borrows money, hires labor and consultants, and has off-farm income. And that's not a bad prospect. In the Agricultural Revolution, as in the earlier Industrial Revolution, ownership of the farm is gradually being separated from its management.

Besides the farms that produce most of the crops and livestock, part-time farms will combine the production of food and off-farm jobs with rural living. This will be a form taken by what was once the family farm.

During the 20th century we flocked to the city. During the 21st century we probably will look to the country as a place to live. The attractions of the country, with its warmer personal relationships, cleaner air and water, more privacy, and greater social stability, will look better and better as the years pass. Increasingly in rural areas, we will be able to provide the social services and utilities that formerly were to be found only in the cities. The Information Revolution, which permitted the flow of knowledge to diverse rural areas, will aid the migration. Practically every industrialized nation now has some kind of rural development program with the objective of making the rural areas a better place to live and work.

During the 20th century we cut down trees; during the 21st century we will plant them. We will put back into trees lands that were deforested to be farmed, lands with slopes too steep for modern farm equipment, or lands too poor to compete with the more productive areas that will be kept in annual crops. Our forests once were considered simply an obstacle to the plow and a source of lumber. In the future they will still be considered a source of lumber but will also be valued for their aesthetic and ecological qualities.

In the past we sought to bring the public domain into private ownership, and to a large measure we succeeded. Henceforward, we shall try to identify and protect the public interest in these privately owned lands. Is the competitive market to be the major, if not the sole, determinant of how these lands are to be used, whether for cropping, grazing, timber, mining, recreation, industry, residential uses, highways, airports, wildlife, watershed protection, or flood prevention? What is the legitimate public interest in the answers to this question, and how is this interest to be expressed? We are in a transitional phase with regard to land policy, and it would take a wise person to see where we might be during the 21st century.

An important institutional challenge during the 21st century will be to manage our system of money and credit so as to achieve a reasonably stable price level, a matter of immense importance to agriculture, especially in a global economy. We are only recently free of a long period of sustained inflation. Will that stability continue? Will it deflate? Or will inflation return?

Maybe during the 21st century no one will be able to buy a pound of butter or a quart of milk or a bushel of wheat anywhere in the United States. We may be on the metric system, with these products selling by the kilogram.

What will be the status of the individual? The 20th century has been called the "Century of the Common Man." The basic rights of full citizenship have been extended, though unevenly, to minority groups and women, who have not previously enjoyed them. Class distinctions have been blurred. These gains are substantial and must be counted, largely to the credit of the generations that advanced them. A floor has been placed over the pit of disaster so that material want, while grievous, is not the problem it once was. During the 21st century these gains will be extended and consolidated. The cautious reader might detect in this stated prospect what may be hope triumphant over uncertainty.

These advances have been achieved by enormously increasing the role of government and by eroding, to a considerable degree, the feeling of self-reliance that once characterized our people. It will be very important, and very difficult, during the 21st century, to ensure that the newfound status accorded to so many of our citizens is used to enhance the role of the individual rather than to deprive him or her of self-reliance. And it will be very important to see that the vast accretions of government power are not exploited by selfish, unwise, ambitious persons. This country achieved its independence by curbing the excesses of powerful leaders and a powerful government; it would be a tragedy if we were to drift back into abuses from which we escaped in years past. It may be that ideologies of the right and left will come to some kind of accommodation.

For the commodity programs in agriculture, the recent trend has been to move away from strong government decision making and in the direction of market orientation. It may be that the high tide of government involvement in the production and pricing of farm products has been reached and that the years ahead will see commodity policies more nearly in keeping with the long tradition of our country.

What of the recent rash of causes that have arisen in agriculture—consumerism, the drive for ecological betterment, concern about genetic modification, and the opposition to food additives? No doubt these drives will continue and some good things will be accomplished. One dares to hope that the excessive zeal that has marked these causes will be blunted. New information that is gradually becoming available should be used to help us check the less admirable aspects of these movements while holding the legitimate gains.

On balance, this is an optimistic outlook. But even if the overall favorable tone should in fact be borne out, we would certainly still have problems. Such is the nature of the human species. Even if all the difficulties that currently are recognized as problems were suddenly solved, a new list of problems would soon develop. Apparently, people have some subjective craving for a balance of hopes and difficulties and are capable of contriving such a balance from almost any combination of the issues that confront them.

So there will be times of exhilaration during the 21st century. And we will hear of wars and rumors of wars, and there will be disasters of various kinds. These events, considered by many to be abnormalities, are to be expected and are likely to be of a magnitude comparable with the pattern of the past.

As a final assessment of the long-term outlook, we quote from what might seem an unusual source: the author of the Book of Genesis. According to that author, the Lord Himself set His rainbow in the heavens and issued this long-term agricultural-outlook statement: "While the earth remaineth, seedtime and harvest, and cold and heat, and summer and winter shall not cease." That forecast has been good for thousands of years. It seems not overly presumptuous to extend it for another century.

References

Avery, Dennis T. 1995. *Saving the Planet with Pesticides and Plastic: The Environmental Triumph of High-Yield Farming.* Indianapolis: Hudson Institute.

Bacon, Francis. 1883. *The Physical and Metaphysical Works of Lord Bacon, Including the Advancement of Learning and Novum Organum.* Joseph Devey (ed). London: Bell. Original work: *Instauratio magna: novum organum scientiarum,* 1620.

Benedict, Murray. 1953. *Farm Policies of the United States, 1750–1950.* New York: Twentieth Century Fund.

Bonnen, James T., and David B. Schweikhardt. 1998. The Future of U.S. Agricultural Policy: Reflections on the Disappearance of the "Farm Problem." *Review of Agricultural Economics* 20(1)(Spring/Summer):2–36.

Carson, Rachel. 1964, 1962. *Silent Spring.* Greenwich, Conn.: Fawcett Publications.

Carver, Thomas Nixon. 1911. *Principles of Economics.* New York: Ginn and Co.

Cochrane, Willard W. 1979. *The Development of American Agriculture.* Minneapolis: University of Minnesota Press.

Cooley, Jack P., and Daniel A. Lass. 1998. Consumer Benefits from Community-Supported Agriculture. *Review of Agricultural Economics* 20(1)(Spring/Summer):227–237.

Cotner, Melvin L. 1976. *Land Use Policy and Agriculture: A National Perspective,* USDA/ERS-670. Washington, D.C.: U.S. Department of Agriculture/Economic Research Service.

Council of Economic Advisers. 1987. *Economic Report of the President.* Washington, D.C.: U.S. Government Printing Office.

Darwin, Charles. 1881. *On the Origin of Species by Means of Natural Selection, or The Preservation of Favored Races in the Struggle for Life.* New York: Appleton. Original work: 1859.

Drache, Hiram. 1964. *The Day of the Bonanza.* Minneapolis, Minn.: Lund Press.

Duram, Leslie A. 1998. Organic Agriculture in the United States: Current Status and Future Regulation. *CHOICES* (2nd quarter):34–38.

Gardner, Bruce L. 1992. Changing Economic Perspectives on the Farm Problem. *Journal of Economic Literature* 30(1)(March):62–101.

Goldsmith, Oliver. 1907. *The Deserted Village.* Illustrated by Stephen Reid. New York: Stokes. Original work: 1770.

Greeley, Horace. 1871. What I Know About Farming. As quoted on page 17 in Earl Ross. 1956. Retardation in Farm Technology Before the Power Age. *Agricultural History* 30 (1)(Jan.):11–18.

Hardin, Garrett J. 1972. *Exploring New Ethics for Survival: The Voyage of the Spaceship Beagle.* Baltimore: Penguin Books.

Hardin, Lowell S. 1998. The Quest for an Evergreen Revolution. Seminar on International Agriculture, Purdue University, West Lafayette, Ind., Nov. 18.

Higgins, F. Hal. 1950. 97 Years of Combining in California. *California Farmer* (March 25):280–281.

Kernan, Joseph E. 1997. Indiana Aquaculture Plan. West Lafayette, Ind.: Cooperative Extension Service, Purdue University.

Keynes, John M. 1936. *The General Theory of Employment, Interest and Money.* New York: Harcourt Brace.

Malthus, Thomas R. 1798. *An Essay on the Principle of Population, As It Affects the Future Improvement of Society. With Remarks on the Speculations of Mr. Godwin, M. Condorcet, and Other Writers.* London: J. Johnson.

Marshall, Alfred. 1890. *Principles of Economics.* London: Macmillan.

Meadows, Donella, et al. 1972. *The Limits to Growth: A Report to the Club of Rome's Project on the Predicament of Mankind.* New York: Universe Books.

Mendel, Gregor. 1965. *Experiments in Plant Hybridisation; Mendel's Original Paper in English Translation, with Commentary and Assessment by Sir Ronald A. Fisher, Together with a Reprint of W. Bateson's Biographical Notice of Mendel.* J.H. Bennett (ed.). Edinburgh: Oliver & Boyd. Original work: *Versuche über Pflanzenhybriden*, 1865.

National Safety Council. 1993. *Accident Facts.* Itasca, Ill.: NSC.

Orden, David, Robert Paarlberg, and Terry Roe. 1999. *Policy Reform in American Agriculture: Analysis and Prognosis.* Chicago: University of Chicago Press.

Paarlberg, Don. 1964. *American Farm Policy.* New York: Wiley.

Paarlberg, Don. 1968. *Great Myths of Economics.* New York: New American Library.

Paarlberg, Don. 1976. Agriculture 200 Years From Now. *Agricultural History* 50(1) (Jan.):303–326.

Paarlberg, Don. 1997. *A Memoir, or a Farm Boy in Washington.* Lafayette, Ind.: Haywood.

Paddock, William, and Paul Paddock. 1967. *Famine 1975! America's Decision: Who Will Survive?* Boston: Little, Brown.

Population Action International. Nov. 1998. *Washington Population Update*. Washington, D.C. Web site: www.populationaction.org.

Projector, Dorothy S. 1964. Survey of Financial Characteristics of Consumers. *Federal Research Bulletin* 50(3)(March):285–293.

Schertz, Lyle P., and Otto Doering III. 1999. *The Making of the 1996 Farm Act*. Ames: Iowa State University Press.

Schultz, T.W. 1964. *Transforming Traditional Agriculture*. New Haven, Conn.: Yale University Press. Paperback: University of Chicago Press.

Schultz, T.W. 1978. *Distortions of Agricultural Incentives*. Bloomington: Indiana University Press.

Sinclair, Upton. 1906. *The Jungle*. New York: Grosset and Dunlop.

Smith, Adam. 1993. *An Inquiry into the Nature and Causes of the Wealth of Nations*. Oxford: Oxford University Press. Original work: 1776.

United Nations, Population Division, Department of Economic and Social Affairs. October 1998. *World Population Estimates and Projections*. New York: United Nations.

United Nations, Statistical Division. 1992. *Handbook of the International Comparison Programme*, Series F, No. 62. New York: United Nations.

U.S. Department of Agriculture. *Agricultural Statistics*, various issues. Washington, D.C.: U.S. Government Printing Office.

U.S. Department of Agriculture, Economic Research Service. *Agricultural Outlook*, various issues. AGO series. Washington, D.C.: USDA/ERS.

U. S. Department of Agriculture, Foreign Agricultural Service. 1953. *United States Farm Products in Foreign Trade*, Statistical Bulletin No. 112. Washington, D.C.: USDA/FAS.

Veblen, Thorstein 1954. *The Engineers and the Price System*. New York: Viking. Original work: 1921.

Von Liebig, Justus. 1840. *Organic Chemistry in Its Applications to Agriculture and Physiology*. Lyon Playfair (ed). London: Taylor and Walton.

Warren, G.F. Jr. 1998. Spectacular Increases in Crop Yields in the United States in the Twentieth Century. *Weed Technology* 12:752–760.

OTHER READINGS

Ahearn, Mary. 1986. *Financial Well-Being of Farm Operators and Their Households*. Agricultural Economics Report 563. Washington, D.C.: USDA/ERS.

Ahearn, Mary. 1990. *The Role of the Farm Household in Agricultural Income and Finance: Situation and Outlook*. USDA/ERS Report AFO-37. May. Washington, D.C.: USDA/ ERS.

McClelland, Peter D. 1997. *Sowing Modernity: America's First Agricultural Revolution*. Ithaca, NY: Cornell University Press.

Paarlberg, Don. 1980. *Issues of the 1980s*. Lincoln: University of Nebraska Press.

Paarlberg, Don. 1988. *Toward a Well-Fed World*. Ames: Iowa State University Press.

Paarlberg, Don. 1999. Obituary for a Farm Program. *CHOICES*, (1st quarter): 33–36.

Randall, C.K., and R.H. Masucci. 1963. Farm and Nonfarm Income Comparisons. *Journal of Farm Economics* 45(2)(May):359–366.

U.S. Department of Agriculture. 1940. Farmers in a Changing World. *Yearbook of Agriculture*. Washington, D.C.: U.S. Government Printing Office.

U.S. Department of Agriculture, Economic Research Service. August 1990. Agricultural Income and Finance, AFO-38. Washington, D.C.: USDA/ERS.

U.S. Department of Agriculture, Economic Research Service. November 1998. *Major Statistical Series of the U.S. Economic Research Service Agricultural Handbook* 691(3). Washington, D.C.: USDA/ERS.

Vasavada, Utpal, and Steve Warmerdam. 1998. Environmental Policy and the WTO: Unresolved Questions. *Agricultural Outlook* AGO-256(Nov.):12–14. Washington, D.C.: USDA/ERS.

Walker, John F., and Harold G. Vater. 1997. *The Rise of Big Government in the United States*. Armonk, N.Y.: M.E. Sharpe.

Index

(Tables are indicated by "t" before the table number following the page number.)

AAA (Agricultural Adjustment Administration), 73
Abundance, 120
Accountability, 132–3
Acts, Freedom to Farm Act, 129
Advantage, comparative, 62
Agenda, farm policy, 128, 130–1
Agent Orange, 40
Agrarian Creed, 6, 21–2
Agribusiness, 67
Agricultural Adjustment Administration (AAA), 73
Agricultural belt, 69–70
Agricultural economics, 61
Agricultural experiment stations, 46
Agricultural fairs, 24
Agricultural policy, 129
Agriculture, commercial, 113–4
Agriculture, industrializing, 130
Agriculture, low-input, 41–2
Aid, purposes of, 105–6
Animal rights, 130
Anthrax, 40
Antibiotics, 37, 131
Apomixis, 139
Archer Daniels Midland, 25
Asia, 80
Avery, Dennis, 42

Bacillus thuringiensis, 45
Bacteria, genetically altered, 44
Baird, John Logie, 51
Bell, Alexander Graham, 51
Bell, David, 75
Benedict, Murray, 74
Benefits, consumer, 117–25
Benefits, human race, questions of, 138

Biology, advancement, 45 t4
Biotechnology, 44–6
Birth rates, 123
Black Watch, 28
Bonanza Farms of the Red River Valley of the North, The, 4, 28
Borlaug, Norman, 43–4, 67, 122
Brown, Lester, 123

Campbell, Alexander, 7
Campbell wheat farms, 28
Carcinogenic foods, 41
Carver, Thomas Nixon, 5
Case, 23
Caterpillar, 23
Cattle, milk, 104
CBK Agronomics, 28
Center for International Forestry Research (CIFOR), 101
Centro Internacional de Agricultura Tropical (CIAT), 101
Centro Internacional de la Papa (CIP), 101
Centro Internacional de Mejoramiento de Maiz y Trigo (CIMMYT), 101
CGIAR (Consultative Group on International Agricultural Research), 99
Changes, biological, 43–9
Chemicals, farming and, 31–42
Chemicals, mistrust, 39–40
Chicago, Ill., 53
Churchill, Winston, 122
CIAT (Centro Internacional de Agricultura Tropical), 101
CIFOR (Center for International Forestry Research), 101

149

CIMMYT (Centro Internacional de Mejoramiento de Maiz y Trigo), 101
CIP (Centro Internacional de la Papa), 101
Coffee, 78, 80
Combine, 4, 23
Commodity compositions, 77–9
Comparative advantage, 60
Competing products, 70
Computers, personal, 57
Concentration and consolidation, 131
Conservation, low input, farming and, 42
Consultative Group on International Agricultural Research (CGIAR), 99
Coolidge, Calvin (Pres.), 72
Corn, 5, 47–8, 104, 122
Corporate farms, 28–9
Cotton, 104
Crick, Francis, 44, 67
Crop research, 103 t8-1

DBCP, 40
DDT, 38
Death rates, 122, 123 t10-2
Deere, John, 4
Demand, 60
Demographic transition, 123, 124 f10-1
Depression, the great, 72–3
Desalination of seawater, 139
"Deserted Village, The," 28
Digestion process, 32
Diminishing returns, 33, 60
Diminishing utility, 60
Disciples of Christ, 7
Diseases, contagious, 37
Diseases of the Horse, 9
DNA, 44–5, 137

Economics
 agricultural, 56, 138
 key principles, 60–1
 terms relating to farms, 61
Economists, 59
Education, 7–8, 99
Efficiency, 24–5
Ehrlic, 123
Electricity, 25–6
Embargos, 76

Embryo transplants, 45
Engineering, plant breeding and, 24
Environmental control, 139
Erie Canal, 3
Erosion, 25
Expansion, 26–7
"Experiments in Plant Hybridization," 43
Exportation, 69–73, 76–8, 81 t7-3
Exports, 72, 79–81

Family, 7–8, 29, 111–2, 119
Farmall tractor, 23
Farm credit, restrictions, 74
Farm Credit Administration (FCA), 74
Farm income issues, 113 t9-1, 113–5
Farming, reduction, 27
Farmland conversion, 120
Farm management, 61
Farm policies, 73–5, 130–1
Farm Progress Show, 24
FCA (Farm Credit Administration), 74
Federal Reserve System, 73
Fertilization, 34 t-1, 130, 139
Firms, biotech, 46
Fish, 48
Fleming, Alexander, 36
Food contamination, 121
Food expenditures, 118 t10-1
Food for peace, 75
Foods, carcinogenic, 41
Foods, illness, 41
Foods, organic, 40–1
"Food will win the war," 72
Ford Foundation, 100
Foreign trade, interferences, 76–7
Freedom to Farm Act, 129
Froelich, John, 22
Fuel alternatives, 139
Fulton, Robert, 3

Garnett, Gwynn, 75
Gates Rubber, 28
General Agreement on Tariffs and Trade (GATT), 76
General Theory of Employment, Interest, and Money, 59

Genetically modified organisms (GMOs), 46, 131
Genetic manipulation, 137
Germplasm, 48–9
Global 2000, Inc., 103
Global conditions, 131, 140
Global Positioning System (GPS), 57, 139
Global production, exports and, 79–81
Globalization, 69–81
GMOs (genetically modified organisims), 46, 131
Goldsmith, Oliver, 28
"Go west, young man," 4
Grain, feed, 62
Great Western Land Company, 28
Greeley, Horace, 4–5
Green Revolution, 44, 123
Gutenberg, Johannes, 51

Hardin, Garrett, 105, 123
Hardin, Lowell, 75
Harrar, George, 75
Health, 130–1
Herbicides, 39
Hill, F.F., 74–5, 106
Hog cholera serum, 48
Hogs, 104
Holstein cows, 62
Hood River, Oregon, 25
Hoover, Herbert (Pres.), 72
Horse teams, 24
Hoxie, C., 4
Humus theory, 31
Hunger, world, 122–5
Hybridization, 43–4
  Mendel's principle of, 47
Hydroponics, 139

ICARDA (International Center for Agricultural Research in the Dry Areas), 101
ICLARM (International Center for Living Aquatic Resources Management), 101
ICRAF (International Centre for Research in Agriforestry), 101
ICRISAT (International Crops Research Institute for the Semi-Arid Tropics), 101
IFPRI (International Food Policy Research Institute), 101
IITA (International Institute of Tropical Agriculture), 101
ILCA (International Livestock Centre for Africa), 102
ILRI (International Livestock Research Institute), 102
Imports, 70, 78, 80, 103–4
Incentives, 60
Industrial growth, 115–6
Information, farms and, 56–7
Information revolution, 51–7
Initiatives, 132–3
Insecticides, 37–9
Insemination, artificial, 45, 48
International Harvester, 23
International Livestock Centre for Africa (ILCA), 102
International Livestock Research Institute (ILRI), 102
International Plant Genetic Resources Institute/International Network for the Improvement of Banana and Plaintain (IPGRI/INIBAP), 102
International Research Network, 75, 99–103, 106–7
International Rice Research Institute(IRRI), 102
International Service for National Agricultural Research (ISNAR), 102
International Water Management Institute (IWMI), 102
Investment capital, 104–6
IPGRI/INIBAP (International Plant Genetic Resources Institute/ International Network for the Improvement of Banana and Plaintain), 102
IRRI (International Rice Research Institute), 102
ISNAR (International Service for National Agricultural Research), 102
Issues, twenty-first century, 137–44
IWMI (International Water Management Institute), 102

Jersey cows, 62
John Deere, 23–4
Jungle, The, 121

Kerosene, 4
Keynes, John Maynard, 59
Khayyam, Omar, 116
Kraus, E. J., 39

Laser technology, 25
Lewis, John L., 127
Liebig, Justus von, 33, 121
Lifeboat strategy, 105
Livestock, 35, 37, 47–8, 104

Malthus, Thomas Robert (Rev.), 125
Management, 59–69
Marconi, Guglielmo, 51
Marketing, 118–9
Market integration, 80
Markets
  decentralization, 53–5
  export, 71
  orientation, 143
  transformation of institution, 55–6
Marshall, Alfred, 61
McCormick, Cyrus, 4
McNamara, Robert M., 75
Meadows, Donella, 105, 123
Meat processing, 53
Mechanization, 5–6, 21–9
Medicine, 9
Mendel, Gregor, 43
Microorganisms, 35–6
Millerites, 7
Morse, Samuel Fairbanks, 51
Müller, Paul, 37–8
Multiphonics, 28
Myers, W.I., 74

NAFTA (North American Free Trade Agreement), 76
National interdependence, 108
Net worth, 111–2

New Holland, 23
Nitrogen, 32–3
North American Free Trade Agreement (NAFTA), 76
Nutrition, 139

Oats, 62
Occupational dangers, 24
O'Neil, Ed, 127
Opportunity costs, 60
Organic Chemistry in Its Applications to Agriculture and Physiology, 32
Organic Foods, 40–1

Paddock brothers, 105, 123
Parity, 127–8
Patterson, James, 4
Payne-Aldrich Act, 71
PBB, 40
PCB, 40
Pearsons, F.A., 73–4
Penicillin, 36
Pest control, 130
Petroleum, 4
Phosphate, 32
Plant breeding, engineering and, 24
Planters, 25
Plow, steel, 4
Poisons, 35
Policies, agricultural, 129
Political issues, 127–33
Pond, George, 61
Population, 123–5, 125 f10-2, 140
Potash, 32–3
Potatoes, 104
Price, determining, 60
Price, one, 60
Price plans, 72
Prices, food, 117
Prices, stabilization, 142
Pricing, food, 60, 119
Processing, decentralization, 54
Production, AAA, 74 t7-2
Production cost and price, 60
Public interest, accommodating, 140

"Qualifications of a Farmer," 64–6
Quality of life, food and, 121

Radio, 55
Railroad, continental, 4–5
REA (Rural Electrification Administration), 25–6
Reaper, 4
Report to the Club of Rome, 105
Research, 46, 67
Reserves, food, 120
Revolution, exporting, 99–108
Rice, 104, 122
Rice Research Institute, International, 122
Rockefeller Foundation, 75, 100
Roundup, 44
Row crop machine, 23
Rubber, 78, 80
Rural development, 27, 141
Rural Electrification Administration (REA), 25–6
Rural life, 6–10

Sarin, 40
Scarseth, George, 33
Schultz, T.W., 106
Selection, artificial, 43
Sheep, 104
Sinclair, Upton, 121
Smith, Adam, 59, 61
Smoot-Hawley Tariff Act, 72–3
Social welfare function, 129
Soil Conservation Service, 120
Soil fertility, 31
Soil management, 139
Sorghum, 104
Soybeans, 5, 62, 104
Stagnation, agricultural, 106–8
Stewardship, 142
Streptomycin, 36
Sugar, 78, 80
Sulfuric acid, 33
Superphosphate, 33
Supply, 60

Tariffs, 71
Taylor, H.C., 61
Technology, 107, 129–30
Thalidomide, 40
Third World, 104, 107–8
Threshing machine, 4
Tissue culture, 45
"Tractored off," 27
Tractors, 5, 22–4
Trade, 70, 71 t7-1, 76–7
Transforming Traditional Agriculture, 106
Transportation, midwest, 70
"Triage" strategy, 105
Tsetse fly, 139
Tull, Jethro, 31
Twenty-first century issues, 137–44
2,4,5-T, 40
2,4-D, 39

U.S. Bureau of Animal Industry, 48
U.S. Department of Agriculture (USDA), 54
USAID (U.S. Agency for International Development), 104–6

Vegetable imports, 80
Von Liebig, Justus, 31–2

Waksman, Selman, 36
WARDA/ADRAO (West Africa Rice Development Association/ Association pour le Developpement de la Rizeculture in Afrique de l'Ouest), 102
Warren, G.F., 61
Warren, G.F., Jr, 62
Water and molasses, 61
Watson, James, 44, 67
Wealth of Nations, the, 59
Weather prediction, 139
West Africa Rice Development Association/Association pour le Developpement de la Rizeculture in Afrique de l'Ouest (WARDA/ADRAO), 102

Wetland preservation, 131
Wheat, 104, 122
Whitney, Eli, 3
Wickard, Claud, 75–6
Wilson, Woodrow (Pres.), 72
Winrock International, 103
Work hours, 35 t3-3

World hunger, 122–5
World War II, 75–6
Wortman, Sterling, 75

Yields, crop, 34 t3-2, 62, 63 t6-1
Yields, effect of, 64 t6-2